SCIENCE AS A CONTACT SPORT

SCIENCE

AS A | INSIDE THE BATTLE TO SAVE EARTH'S CLIMATE

CONTACT SPORT

STEPHEN H. SCHNEIDER

 NATIONAL GEOGRAPHIC

WASHINGTON, D.C.

To my grandson Nikolai Cherba, in the hope that our efforts
will make the world you inherit from us more resilient

Published by the National Geographic Society

1145 17th Street N.W., Washington, D.C. 20036

ISBN: 978-1-4262-0540-8

Library of Congress Cataloging-in-Publication Data
Schneider, Stephen H.
 Science as a contact sport : inside the battle to save Earth's climate / Stephen H. Schneider.
  p. cm.
 Includes bibliographical references and index.
 ISBN 978-1-4262-0540-8
 1. Greenhouse effect, Atmospheric. 2. Global environmental change. 3. Climatic changes--Government policy--International cooperation. 4. Global temperature changes. I. Title.
 QC912.3.S33 2010
 363.738'74--dc22

                    2009029206

Illustration Credits: Page 13: Data courtesy NOAA, Page 18: Image courtesy of the author, Page 44: Data courtesy of NOAA, Page 135: IPCC, Page 190: From Joel B. Smith et al., "Assessing Dangerous Climate Change through an Update of the Intergovernmental Panel on Climate Change (IPCC)" "Reasons for Concern," Proceedings of the National Academy of Sciences 106 (2009): 4133-37, Page 247: Image courtesy of the author.

The National Geographic Society is one of the world's largest nonprofit scientific and educational organizations. Founded in 1888 to "increase and diffuse geographic knowledge," the Society works to inspire people to care about the planet. It reaches more than 325 million people worldwide each month through its official journal, *National Geographic,* and other magazines; National Geographic Channel; television documentaries; music; radio; films; books; DVDs; maps; exhibitions; school publishing programs; interactive media; and merchandise. National Geographic has funded more than 9,000 scientific research, conservation and exploration projects and supports an education program combating geographic illiteracy. For more information, visit nationalgeographic.com.

For more information, please call 1-800-NGS LINE (647-5463) or write to the following address:

National Geographic Society
1145 17th Street N.W.
Washington, D.C. 20036-4688 U.S.A.

Visit us online at www.nationalgeographic.com

For information about special discounts for bulk purchases, please contact
National Geographic Books Special Sales: ngspecsales@ngs.org

For rights or permissions inquiries, please contact
National Geographic Books Subsidiary Rights: ngbookrights@ngs.org

Printed in the United States of America.

Interior design: Cameron Zotter

09/QWF-ML/1

CONTENTS

FOREWORD

STEPHEN SCHNEIDER HAS SEEN more of climate change politics and climate change science than almost anyone alive, and he's been so effective at countering the climate skeptics and lobbyists that he's become a special target of their campaign to discredit leading scientists. Such an experience would embitter many, but somehow Stephen Schneider has retained his good humor and balance throughout. Having been at the forefront of the push to have climate change addressed during the dark years of the Bush Administration, this itself is a singular achievement. But Steve has done far more than this, for he is a practicing climate scientist who has contributed significantly, and his passion for science has inspired many, including ourselves, to join the campaign to address climate change.

I first met Stephen Schneider in Japan nearly a decade ago, at a conference on extinction threats. His words on the danger of a changing climate to biodiversity hit like a thunderbolt, and from then on I was convinced of the truly dire nature of the threat that climate change is to our planet. His presentation was clear, packed with information, and funny. It was the last thing I expected from a great man addressing a serious topic, but I soon learned that one of Steve's greatest assets is to bring humor to overly serious debates. Indeed, it seems likely that Steve would have had a great career in stand-up comedy if he had not devoted his life to science.

Google Stephen Schneider and you will discover why this gift for humor is so precious. Blogs still use his work from the early 1970s on a possible future ice age to discredit his later climate science. Yet at the time his musings on a possible cooling trend were entirely mainstream. The skeptics seem to believe that science cannot move on and progress. What they have found in Stephen Schneider is a scientist who can and does progress, and who is articulate in his exposure of their deceits.

Each of us tries to contribute to combatting climate change in our own way. In his latest book, Stephen Schneider offers us a history of a bruising time in the politics of climate change and uses those lessons of history to build a pathway forward. It is a time now thankfully behind us, and nobody has contributed more to that extraordinary outcome than Steve himself. He well deserves the Nobel Peace Prize shared with Al Gore and Steve's many colleagues in the IPCC.

Tim Flannery
Chairman, Copenhagen Climate Council

INTRODUCTION: THE GLOBAL PLAYING FIELD

THREE AND A HALF DECADES AGO, a few dozen scientists around the globe appeared in front of many of the world's legislatures, including the U.S. Congress, warning of potential climate changes caused by humans using the atmosphere as a sewer to dump our tailpipe and smokestack wastes. Solutions like halting deforestation, protecting ecosystems, using solar power, requiring better building insulation, and developing fuel-efficient cars were widely proposed.

Sound familiar? We hear a lot of those same terms now. But a generation and a half later we still have no significant policy response at a global level to this mounting threat. How did that happen? Who blocked the efforts? Who tried to help? What can we learn from our failures? Do we still have time to avoid dangerous consequences? Those are among the principal issues that we'll explore in these pages.

When I stepped into the science arena four decades ago, the term "global warming" was virtually unheard of. Only a handful of young people were majoring in newly created environmental studies programs. Physicists and meteorologists were running data analyses on mainframe computers using stacks of punch cards, which often took days to type out, card by card—and then hours to process at the computer center. We would drop off the boxes of cards one day and pick them up the next with the results. Color graphics were scarce. Color monitors? In your dreams. For graphics, a big microfilm reader was considered state of the art, and long lines of users waited their turn. Smart computer programmers kept a back-up deck of punch cards to hedge against the dreaded operator who, in a rush to load all the boxes of cards, dropped a box, leaving your days of work scrambled on the floor. Today, my multifunction high-tech watch can calculate as fast as some of the "mainframes" of yesteryear.

Back then each branch of the sciences inhabited its own discrete domain—the atmospheric researchers didn't talk much to the computer program designers or the oceanographers, let alone the urban designers, ecologists, economists, or sociologists. The concept of an interconnected, interdependent whole Earth system hadn't evolved beyond a few visionary thinkers. Despite that, even in 1970, at the height of the Cold War, atmospheric scientists, at least, communicated internationally. This came about not through some vision of globalization or glasnost decades ahead of schedule, but simply for practical self-interest. Atmospheric scientists everywhere had a common cause: We needed expensive satellites, balloons, ships, and computers from our governments to do our work. International cooperation in data sharing reduced the cost to individual nations.

In the summer of 1971, for instance, I served as the rapporteur (report writer) for an international meeting of two dozen senior scientists in Stockholm. They spent a month addressing inadvertent climate modification, for the first time at a deep level. Not only was the term "global warming" not in circulation yet, but at that time we didn't know whether warming from carbon dioxide (CO_2) or cooling—from the air pollution haze in cities or the burning of biomass (agricultural wastes or forests burned for land clearing) in rural areas—caused a bigger effect.

Because of the Vietnam War, the United States was not very popular in Sweden. Yet we and the Swedes and other Europeans had a common bond in our desire to understand the planetary threats—and to guard against the Soviets and allied states who had imprisoned most political dissidents. Despite Vietnam, the Cold War, the Stalinist hangover, and many sharp international disagreements, the scientific community was able to coalesce around a theme of common interest in our collective life support system and get on with the work of understanding and protecting it. To this day, I am proud of that early beginning in global cooperation—the most essential factor in facing the climate change threat.

Interest in global science was spurred by the space race of the 1960s. A Soviet cosmonaut had snapped the first photo of Earth from space in August 1961. The dramatic images of our blue-green planet captured by NASA's space explorations through the decade had an enormous impact on how people—not just scientists—viewed the world. Gradually, we began to recognize the fragility of Spaceship Earth and its ecosystems, and since the early 1970s, the science community has been increasingly wrestling with issues of climate change as a global problem.

Those early photos of Earth have grown in sophistication. In July 2008 a team from NASA's Jet Propulsion Laboratory published global satellite maps of the key greenhouse gas, carbon dioxide, in Earth's mid-troposphere, an area about 8 kilometers (5 miles) above Earth. The map reveals how carbon dioxide is distributed in Earth's atmosphere and moves around our world. The bright green, red, and yellow swirls create an undeniable depiction of the global effect of this greenhouse gas. The distribution of CO_2 respects no international boundaries. [1]

Today, climate change is acknowledged by most climatological experts around the world. Some have replaced the term "global warming" with "global heating" or "the global heat trap," or simply "climate disruption," to indicate that humans are contributing to what is occurring. Many scientists, in their endless striving to prove dispassionate objectivity, call it "anthropogenic climate change"—an accurate phrase, but not a favorite of newspaper headline writers and TV anchors.

This acknowledgment of global concern has been achieved through surmounting numerous obstacles along the way. Policymakers, lobbyists, business leaders, and extreme skeptics have struggled mightily to steer public opinion—and the funds associated with it—in their preferred directions. Most mainstream scientists have fought back with the weapons at their disposal: methods of truth seeking such as peer review, responsible reporting of research data, best practice theory, international cooperation, and cautious calls for policy consideration. The battle is by no means won. The world needs all our combined strengths to cope with the dangerous climate impacts already under way, much less prevent far more damaging climate change 20 or more years from now.

Why haven't we made more progress in stopping the potential disruption from greenhouse gas emissions and deforestation, which increases warming of the atmosphere and ice melting? Why do alternative sources of energy equal only a small fraction of the polluting fossil fuels and questionable oil drilling in sensitive environments? Have government policies contributed to record profits for oil companies? What has been going on for the 40 years since this problem was first demonstrated to government officials all over the world?

The answers are both simple and complicated. The simple can be summed up in five easy pieces: ignorance, greed, denial, tribalism, and short-term thinking. Let's face it—with so many billions of people to feed, house, and help to be productive, we focus on the immediate, not what is sustainable over decades to centuries. The complicated aspects will require most of the chapters in this book to answer. But answer we must to prevent most of the looming danger. We must first see what held back beneficial changes from being implemented and then fashion strategies to overcome these constraints. As the saying goes: Those who don't learn the lessons of history are doomed to repeat them.

My memories and convictions are those of a climatologist who has worked in the computer lab and sometimes in the field, testified before legislative committees, and worked with businesses, universities, public service clubs, and even elementary schools. Although objectivity and dispassionate presentation of empirical facts and plausible theories are required of the scientist in a professional capacity, in this book I am writing from my personal perch in the battlefields. I tell the truth as I have seen it and as I remember it. I believe most recounted stories and quotes are in context and close to the original, but I make no claim that my memory is perfect; however, I have sent early drafts of this manuscript to many others who were there and asked them to confirm my memories. In most cases we agreed, but sometimes others remembered things a bit differently. So all I can claim is that as an overall personal history, this is to the best of my ability a fair accounting of events, and I also have tried to verify it by seeking the opinions of others to a considerable extent. Nonetheless, while I believe all quotes from memory—mine or for others—should be considered seriously as part of the narrative, I do not ask that they be taken literally, as if they were recorded and transcribed.

I've been here on the ground, in the trenches, for my entire career. I'm still at it, and the battle, while looking more winnable these days, is still not a done deal.

THE PRESENT STATE OF CLIMATE CHANGE AWARENESS

I believe in a hopeful environmental future for the world if we proactively address the challenges of climate change and their strong connection to the sustainable development of the not yet well-developed countries. We can overcome the political inertia that has delayed our response here in the United States as well as in many other countries. Consumption and population are the main drivers of environmental stresses, and both need to be managed for sustainability. As peoples' health and nutrition improves and as women become more educated and have the opportunity to choose their family size, population growth rates have come down considerably. While it may have its social critics, China's one-child policy has been a major factor in reducing the country's long-term potential emissions. The world's population is still likely to grow from our current 6.8 billion people to 8 or 9 billion people by century's end, but the older projections of more than 12 billion now seem less likely. However, if all these people expect to attain the consumption levels of a typical American today and at the same time derive their energy mostly from fossil fuels, then major climate change is virtually inevitable.

What are we doing? Factories that make the gas-guzzling cars and SUVs are being forced to retool—not only by environmental regulations but by market forces such as the price of oil. People's consumption, which drives emissions, is being tempered by the realities of the economy and the environment. The coal industry has some hard thinking to do about its place in an administration that supports solar and wind power investment, sensible alternative biofuels, some well-designed nuclear expansion, smarter electrical grids, and geothermal exploration. Those ideologically opposed to having the government directly involved in technology development might want to look again at the historical record. Were jet planes, computers, electronics, autos, and coal and nuclear power invented without major investments by governments? Hardly. The private sector can mass-produce or build a product better and cheaper than governments, I agree, but rarely do they initiate new ways to thrive without substantial governmental priming of the pump. When major *Fortune* 500 corporations like General Electric, Wal-Mart, Duke Energy, and Pacific Gas and Electric say "we need a climate policy," a bipartisan effort to establish that "thought leadership" becomes a real possibility. But it won't happen fast enough or forcefully enough without more than voluntary actions by these corporate leaders.

The climatic future is already upon us, at least in violent spurts. In 2003 an unprecedented heat wave in Europe killed more than 50,000 people (mostly elderly), because no one realized how vulnerable they were and thus few plans had been formulated to help them adapt to such an unexpected, unprecedented extreme event. The heat wave solidified the movement that was already building in Europe to support climate policy. In August 2005 we lived through the disaster of Hurricane Katrina, which will remain forever a legacy of the incompetent policies of the Bush Administration's emergency services agencies.

Climate change experts have long projected that the strongest hurricanes will intensify as they move over warmed oceans, although details remain controversial. As predicted, wildfires in the western United States have dramatically increased: Five times more area has burned in wildfires in the past 30 years than ever recorded before. The sea levels are rising, high mountain and polar glaciers are melting, and the Arctic sea ice is rapidly thinning all year long and increasingly disappearing in summer—all predicted decades ago and now playing out at rates as fast or faster than was projected. When I am asked what is the difference between our early, carefully hedged warnings 30 years ago and now, I often say, "What has changed is not the basic science so much as the fact that nature is cooperating with theory."

How the U.S. government answers the challenges of climate change will depend in part on the leadership of the majority Democratic Party and its efforts to reach across the aisle (and across the oceans) and also on the personal commitment of individual citizens and businesses. If only President Obama and former rival Senator John McCain—an early supporter of climate action—could unite in showing leadership from one end of Pennsylvania Avenue to the other, we might at last achieve meaningful climate policy.

U.S. polls indicate that people finally recognize climate change as a pressing national issue—although even those who believe global warming is real and probably caused largely by humans still rank climate change lower in priority than the economy or national security. If we cannot change that attitude soon, calamities such as Katrina and the European heat wave will seem like mild omens in the decades ahead. Actions take consensus and commitment, and we are three decades late. Further delay may tip us into conditions that will be very difficult to reverse, including melted ice sheets, redrawn coastlines, more dangerous wildfires, increased killer air pollution episodes, and species driven to extinction.

Former Vice President Al Gore hit a grand-slam homer for climate change awareness when he produced his Academy Award–winning documentary film, *An Inconvenient Truth*. But it wasn't his first time at bat. He began his environmental career more than two decades earlier.

In October 2007 several of my fellow Stanford scientists and I joined the former Vice President on the podium for a press conference, announcing the Intergovernmental Panel on Climate Change (IPCC) as a co-recipient with Gore of the Nobel Peace Prize for its work on global warming. The IPCC had just released most of its Fourth Assessment Report, which assigned "very high confidence" (at least a 9 out of 10 chance) to serious impacts due to human-caused global warming.

The IPCC report, entitled *Climate Change 2007*, encompassed all regions of the world, integrating data and analyses compiled from years of work by the scientific community in published peer-reviewed journals. Yet its statements differ surprisingly little from the testimony I gave at one of the first set of contentious hearings on climate change in the U.S. House of Representatives, conducted by the then junior congressman from Tennessee, Albert Gore, in 1981. The situation has only gotten worse. Not only are we experiencing much of what was predicted, but it is often happening faster and wreaking greater havoc.

In a world of rapid climate change, we all have to develop risk assessment skills in order to determine the best courses of action given the global costs and benefits. How did we get to this point of staring cataclysmic changes in the face? How can we move effectively toward the long-delayed imperatives that lie ahead? The first step is to examine the past and draw the lessons from it to guide our way to a hopeful future.

SMOKE ON THE HORIZON

1

SCIENTISTS HAVE BEEN DISCUSSING the risks of human-induced climate change for decades now. They've testified before Congress, presented research at scientific conventions, engaged with the media (for good or ill), debated with their peers, and alerted policymakers around the globe. Yet half a century after serious scientific concerns first surfaced, the world's governments are still far from a meaningful accord on how to adapt to the unavoidable consequences of climate change or mitigate changes that we can't easily adapt to.

Protection from unacceptably severe impacts is high on the list of priorities for those scientists and citizens who have been paying attention, but the honest admission of the scientific community that we can't predict most of the details with absolute certainty leads many people to downplay the threat. Much of my work as a scientist these days focuses on getting the message out to the folks who need to act—*now*—to reduce harmful activities and adapt to the range of possible outcomes that can't be avoided over the next 30 years. It's not only the political and economic leaders that need to hear this message. It's all of you, too, in both hemispheres and every latitude. If you live on Earth, you—and your posterity—have a stake in its future.

As my friends and foes know, I've never been shy about airing my views—even when my career was threatened. I'm not about to abandon that lifelong practice now. I want to recount the story of how climate scientists gradually formed a strong consensus that human activity has produced potentially dangerous changes in Earth's climate. Along the way, I'll unveil how ignorance and duplicity have inhibited the world community from implementing solutions sooner. I write from my own personal experience in the trenches of the climate wars since 1970.

Long before my generation arrived on the scene, scientists had begun looking into the level of carbon dioxide in Earth's atmosphere. In 1824, the French mathematician and physicist Joseph Fourier first hypothesized that CO_2 might be implicated in the state of Earth's climate. As early as 1896, the Swedish chemist Svante Arrhenius published the first theoretical study explaining how carbon dioxide and natural water vapor trapped the sun's heat in the atmosphere, linking CO_2 even at that time to the burning of fossil fuels.

Both Arrhenius and Guy S. Callendar, an English engineer writing in 1938, believed that the warming of the planet would have a beneficial effect, increasing harvests and raising temperatures to make frigid regions more hospitable to human residents. That has been a longstanding hope for some northern countries, including the former Soviet Union, which in the 1960s commissioned a pamphlet, "Man Versus Nature," calling for deliberate modification of the atmosphere to warm the Poles. Not surprisingly, they largely ignored potential unexpected consequences of monkeying around with the global weather machine.

The idea that the atmosphere was like a glass-paned greenhouse keeping the surface temperatures of Earth warmer was the analogy used in 1937 by Glenn Trewartha, a University of Wisconsin geographer who helped coin the term "greenhouse effect." Oceanographer Roger Revelle and his Austrian colleague Hans Suess, a chemist and nuclear physicist, collaborated on a paper published in 1957 in which they concluded in an oft-cited statement that "human beings are now carrying out a large-scale geophysical experiment" by emitting greenhouse gases into the atmosphere, without accurate measures of the levels of carbon dioxide in the atmosphere and without knowing how much those levels were changing.[1]

The United States and other Western nations after World War II pursued policies of everything "bigger and better"—exponential growth of the economy driven by more cars, houses, appliances, fertilizers, pesticides, and so forth, without serious accountability of accumulating side effects. For decades technological and economic hubris had led us to believe we could master nature and get rich simultaneously.

Yet Rachel Carson's seminal 1962 book *Silent Spring* was the most notable of the early genre of warnings that human technological development could disturb the functioning of nature. The eventual ban on DDT in most rich countries is credited with saving countless birds and eventually lifting such species as the bald eagle and the peregrine falcon off the endangered species list. It also serves as a positive example that when we perceive a threat to nature, even one that requires costly actions to prevent, we can mobilize public support to implement conservation policies. The ban opened up markets for substitute products and became an incentive for researchers to invent chemicals that could do the job with fewer environmental side effects.

Revelle first pointed out that the oceans were not likely to absorb all of the carbon dioxide emitted into the atmosphere by the burning of fossil fuels. He was a pioneer of interdisciplinary studies, and Revelle College at the University of California–San Diego was his honorary legacy. Once when I was driving in La Jolla with him on the way to a meeting, in his modest, aging Japanese compact sedan, my curiosity got the better of me. I asked him, "Roger, what did you do to get the Revelle College named after you—it was you, I presume?"

"Right," he answered. After a minute, he smiled and said, "Well, I suppose they thought that talking the Navy into donating the land on the hill in La Jolla to the University of California to build UCSD was probably enough to get their attention."

Revelle was one of the founders of the International Geophysical Year (IGY), which was designated by the scientific community for 1957–1958 as an awareness campaign to encourage research into the physical state of the planet. In 1958 he was the director of the Scripps Institution of Oceanography when he asked a young Caltech geochemist and oceanographer named Charles David Keeling to join his team in La Jolla. Dave Keeling had built the first instrument capable of accurately measuring the level of carbon dioxide in atmospheric samples, and it had already shown that the level had risen since the 19th century.

If scientists wanted to track continuing levels accurately, it was imperative to start right away, in a location where the sampling of the purest air and most stable results could be obtained. Revelle and Harry Wexler of the U.S. Weather Bureau obtained funding for Keeling to set up his new instrument at a base two miles above sea level on Mauna Loa volcano in the Big Island of Hawaii, a location in the Pacific that was freer of local influence from continental landmasses than just about anywhere else that was convenient for a laboratory.

Keeling began recording his atmospheric measurements in March 1958 and successfully graphed seasonal variations in CO_2 levels in the first years. By 1961 his data clearly showed a steady annual increase in carbon dioxide levels in the atmosphere, beyond the seasonal oscillations, and he correlated the trend to humanity's burning of fossil fuels. The graph that demonstrates this increase is called the Keeling Curve. The ongoing measurements at Mauna Loa document a rise in atmospheric CO_2 from 315 parts per million (ppm) in 1958 to 385 ppm in 2009.

Atmospheric CO$_2$ concentration (ppm)

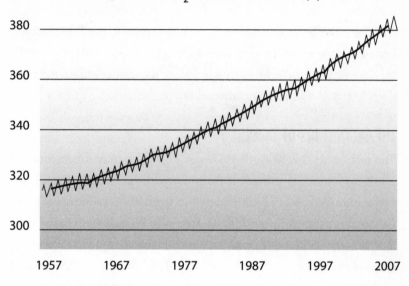

*Dr. Charles David Keeling's famous curve shows increasing
levels of CO$_2$ in the atmosphere since 1958.*

When I met Dave Keeling in 1972, I expected to meet an activist, since he was by then famous for his unique curve and his warnings that climate change could be a significant issue. But instead, he was a modest scientist who was a stickler for accuracy in measurements and careful in interpretation of his data.

In November 1972, Jerome Namias, a well-known meteorologist who specialized in long-range forecasts, called a meeting on the new excitement in climatology at Scripps. I was invited, at the age of 27, to lecture on climate modeling to these famous senior scientists. Dave, a chemist, stayed through the entire three-day meeting, attending every equation-laden talk, many not in his field. He wanted to learn, and by example he was an important mentor for me. What I learned from Dave Keeling was to keep the science impeccable, work on important problems, and tell people simply what you know when they want to know it.

When the Scripps Institution, in memory of Dave Keeling, celebrated the 50th anniversary of the Keeling Curve in 2008, its website message expressed the hope that in future we may celebrate the news that the curve is heading downward at long last.

THE BEGINNINGS OF CHANGE

The first Earth Day in April 1970 was a celebration of our planet and a call for action to protect it from dangers that were being identified more clearly every year. Essentially a grassroots movement, Earth Day was organized in the United States by Wisconsin Senator Gaylord Nelson and a young activist, Denis Hayes, as a national demonstration of concern for the environment. It followed along in the spirit of the massive anti-Vietnam War demonstrations rocking the country at that time. All across America, particularly on college campuses, the events brought hundreds of thousands of people together in a public recognition of the urgent need to take steps immediately to save Mother Earth. Like groundhogs emerging from burrows, on April 22 we looked around and blinked in surprise at the number of people who came out to participate. The individuals and local groups supporting environmental advocacy realized in a single day that they were not alone. There was power in numbers. Earth Day 1970 delivered a whopper of a message.

The consequences of treating our planet as a garbage dump—greenhouse gases were not a primary focus at that time—were brought into public focus at a watershed moment in history. The cultural values of America were in the throes of a major paradigm shift that began in the mid-1960s, when the Beatles knocked Elvis off the pop charts and the U.S. military escalated its role in Vietnam. The conformity of the 1950s gave way to open, experiential challenge in all things, including scientific research. The younger generation broke free of their parents' mantras of "America first" and knee-jerk patriotism. The military draft also motivated both the draft-age men and many of their parents to be alert to Bob Dylan's message: "The times they are a-changin'." Without Vietnam, I'd guess that the anti-status quo environmental movement might have lain dormant for quite a while longer.

I was a graduate student at Columbia University in the late 1960s, working under C. K. "John" Chu in mechanical engineering. John Chu, well known today as a pioneer in computational mathematics, was also interested in how science made a difference in the world, not just in science for its own sake. He once said to me, "If what you do doesn't make a difference in the world, what good is it?" At the same time, as a Courant Institute of Mathematical Sciences–trained numerical analyst, his methods were highly rigorous.

I selected John Chu as a thesis adviser, not because I had passionate interest in partial differential equations, plasma physics, or numerical fluid mechanics, but because I liked his style. I didn't want to discover something just for its own sake. I wanted what I did to make a difference in the world too. What I learned from John went far beyond working with ionized gases and collisions of particles in the hope of creating abundant, low pollution energy, which was the motivation for much of the plasma physics work at Columbia. His style—his belief that science should make a difference—I took as permission several years later to be open to the new world of climate science, even if I was still paid to do plasma physics. By that time I knew that you could learn a field by going to seminars. If I listened carefully to the most knowledgeable mentors at the meetings, I could assess where the cutting edge of a field was even before I had a deep understanding of it. To achieve the latter takes a lot more than attending seminars.

In early 1970, just prior to Earth Day, meteorologist Joe Smagorinsky, founder of the Geophysical Fluid Dynamics Laboratory, came up from Princeton and gave a talk about climate and weather modeling at the plasma physics colloquium at Columbia. By this time I knew something about numerical modeling of complex fluid systems, and I really enjoyed his talk. It reminded me of my fascination with hurricanes when I was a kid, when I used to go up to the attic to track the wind-speed meter on the anemometer my dad bought us. Everybody else in the neighborhood was saying, "God, the trees are going to hit the house," but I was up there because it was exciting.

"Smag," as everybody called him, gave this talk about geophysical fluid dynamics, yet another application of basic principles of fluid mechanics, which studies how liquids and gases move under various forces, in his case from Earth and its atmosphere. From my work on plasma shock tubes with John Chu, I knew how to calculate magneto-hydro-dynamic shocks at 20,000 times the speed of sound, but I didn't know anything about low-speed air currents or oceanic currents moving on a rotating sphere like Earth. I was creating one-and-a-half–dimensional models of ionized gases, but for detailed weather and climate calculations, you need three-dimensional models and you have to include planetary rotation. The subject started to interest me scientifically and as a social need, to the point that I thought I should do a postdoctoral study on climate.

I searched the Columbia course catalog and found only a few classes on climate and atmospheres. I sat in on a climatology course taught by geographer John Oliver and found it fascinating. It was very qualitative, relative to what I was doing, but at the same time I learned all the basic concepts in climatology from a geographer's perspective. The homework assignments involved plotting graphs of temperature between land and oceans, or around overheated cities. I didn't consider this very rigorous, having come from a plasma physics and applied math background, where calculations based on complex theories were the norm. Nevertheless, the subject matter was really interesting.

Wally Broecker and Arnold Gordon, whose oceanography class I also sat in on, were more rigorous. They used so-called box modeling—breaking up the ocean into a top layer and a deep layer, for example, and having terms for the flow of chemicals and water between the boxes to simulate as simply as possible the processes in the oceans.

But what really hooked me was S. Ichtiaque Rasool's graduate seminar on planetary atmosphere. Why was Mars cold, Venus hot? he asked. Why did Earth not become like Venus? How did a cold trap work? How does the greenhouse effect work? Rasool was the number two guy at NASA's Goddard Institute for Space Studies (GISS) in Manhattan. Although a satellite expert, he had widespread planetary science interests, and he had a phenomenal sense of what was important. As a sign of his shrewdness, he had just hired a young postdoc to do radiative transfer calculations of planetary atmospheres, James Hansen, who has since become an authority on climate change and an international icon of the need to protect the planet.

GISS, also known as the Institute for Astrophysics, was run by Robert Jastrow. Everybody called it the "Institute for Jastrowphysics," in part because he was a tyrant who demanded everything be done exactly the way he wanted when he wanted it. He would call subordinates on Saturday afternoon and say, "I need you down here right now because I have an idea!" and if people responded, "I'm in the middle of my kid's birthday party," he would snarl, "You want a job?!" He was not a popular boss. But the institute was an oasis of a NASA lab, located right over Tom's Restaurant at Broadway and 112th Street, made famous by Jerry Seinfeld decades later.

We used to have exciting meetings there. Jim Hansen was studying planetary atmospheres. Rasool brought me in as a part-time grad student while I was finishing my thesis for Chu. But the climate modeling field was still in its infancy. When the environmental biologist Barry Commoner came to Columbia, he gave an Earth Day speech

Steve Schneider, Jim Hansen and S. Ichtiaque Rasool, circa 1971

saying that CO_2 is going to warm Earth, and aerosol particles are going to cool Earth, and we don't know which is going to win. After I heard Commoner speak, I asked Rasool, "Who is studying this CO_2-aerosol problem? I mean, is the Earth going to warm; are we going to cool?"

Rasool said, "Why don't you work for me this summer on the problem? I'll give you some computer codes we have lying around the building, and you can modify them." He handed me a batch of equations and computer codes, the primary one of which was an infrared radiative transfer code that calculated how greenhouse gases absorbed some of Earth's radiant heat. So once again I used my skills in numerical methods to solve equations I didn't understand.

The open question centered on the effect of aerosols in the atmosphere. Any particles suspended in a gas are called aerosols. An aerosol spray can draws liquid from inside the can and with compressed gas inside, propels it out through a nozzle to aerosolize it—that is, make it into droplets suspended in a gas. Being liquid, they evaporate nearly immediately. But at a much grander scale the particles that make up the hazes and smoke from industrial pollution or agricultural or desert dust can blow high into the atmosphere and last for weeks. They can spread a thousand miles downwind of the source and affect the amount of sunlight that is absorbed and reflected in the atmosphere. Since most aerosols are lighter in color than the surfaces they float over, they reflect away sunlight—thus they cool the climate.

Climatologist Reid Bryson from the University of Wisconsin had become a controversial figure by suggesting that overgrazing of vegetation cover in India resulted in blowing dust that cooled the climate and suppressed the life-giving rains of the Indian monsoons. We focused on the aerosols caused by industrial pollution and arrived at similar conclusions to Bryson. We thus became temporary allies in the warming-versus-cooling battles to come shortly after Rasool's and my paper was published.

However, I needed a simple way to calculate the aerosol effect. Over lunch with Jim Hansen one day, I explained my predicament. He said that Carl Sagan and his former student Jim Pollack had come up with a pretty quick solution to aerosol-scattering problems. I used that idea.

The driving assumptions of our paper were that the aerosols were global and the greenhouse gases—CO_2 only—were also global in extent. The model that I was given by Rasool had no stratosphere— the layer of the atmosphere above the turbulent troposphere, where all the water clouds are and the mixing takes place. The stratosphere is where high-flying jets cruise and where Earth's life-protecting ozone layer predominately exists. At that time I had no idea that leaving the stratosphere out of our greenhouse effect computer calculations was such a large error. I didn't learn until later that running the model without a stratosphere was going to cut in half the climate's sensitivity to CO_2 increases. As a result, we only calculated about 0.7 or 0.8 degree Celsius (about 1.4 degrees Fahrenheit) warming if the amount of CO_2 doubled. In contrast, when we ran the model with aerosols as if they were everywhere, our global CO_2 effect was swamped by the global aerosol effect, and we predicted cooling of 3 to 5 degrees Celsius (5.4 to 9 degrees Fahrenheit) by the year 2100. Rasool, quoting the controversial work of climate modelers Mikhail Budyko in Russia and William Sellers at the University of Arizona, even wrote at the end of our paper that 5 degrees Celsius cooling could trigger an ice age!

Rasool's throwaway line was going to come back to haunt me. Nearly 40 years later polemicists are still trying to ridicule me for "predicting an ice age" then and global warming now. Conservative columnist George Will even referred to me in a 1996 op-ed piece in the *Washington Post* as "an environmentalist for all temperatures."[2]

Lay commentators like Will clearly don't understand that this is how science progresses, continuously correcting its conclusions based on new research. You build your case on existing literature, explain what original findings or ideas you are adding, state your assumptions transparently, calculate the consequences as if those assumptions were true, and then redo your calculations after debating with your colleagues, learning more, and reading the latest literature. In science, we are proud of getting the wrong answer for the right reasons, and we're especially proud if we ourselves are the first to correct it.[3]

Later, with Tzvi Gal-Chen, my colleague at the National Center for Atmospheric Research (NCAR), we rebuilt those Budyko-Sellers models, did climate stability calculations with them, and showed they were less unstable than earlier believed. We ended up significantly changing the climate stability debate—but it still isn't fully settled.

The paper Rasool and I wrote, entitled "Atmospheric Carbon Dioxide and Aerosols: Effects of Large Increases on Global Climate," was published in July 1971 in *Science* magazine.[4] In a way, I don't deserve much credit for that paper even if I did all the calculations, because I did not make the key assumptions. I earned that credit later, when our paper became a cause célèbre. Rasool asked me to go out and give all the talks defending it, because people were really trying to shoot it down, a common practice for new claims, especially ones as controversial as this.

CLIMATE SCIENCE LIGHTS A FIRE

When I was working for Rasool over at Goddard Institute for Space Studies, I would run my plasma physics thesis code on the GISS computer. I'd enter the box of cards personally in the remote card reader room, and since it was a short program, I would hang around the room and watch the IBM mainframe churning on the big weather forecast model they were continuously running—a GCM, or general circulation model, that tracked atmospheric flow. I had now pretty much made up my mind that I was going to switch into climate science. I was really getting interested in it, and the work was critical for Earth's environmental health. The inherent physics, the fact I could experience the weather every day, and the societal application of the climate problem created a perfect synergism for me in which my environmental interests meshed with my training in applied physics. I could sit down at the key punch, type up a box of cards, and hold in my hands the capacity to simulate Earth's climate, polluted or not. Of course, little did I know how ill informed these models could be, built on the often arbitrary assumptions of four decades ago.

Nonetheless, I was drawn to the power of the idea. We could actually simulate Earth's temperature and then pollute the model in order to figure out what might happen before we had polluted the actual planet. I might even have some positive influence on policy one day, which was something I'd always wanted to do.

I'd had a taste of policymaking after the student riots at Columbia in 1968, when, as a grad student in engineering, I was elected to negotiate with the trustees to create an academic senate. I became a key member of the student negotiating team working with the administration and trustees to add some democracy to the running of the university. What I learned through that tumultuous time has been critical to my efforts as an advocate for climate change awareness. You needed to be credible, open to coalitions of strange bedfellows, and able to negotiate very hard and persistently to have a chance to implement outside-of-the-box changes. Over many decades, I've tried to use this experience to forge coalitions of engineers, environmentalists, scientists, journalists, citizens' groups, and policy wonks to coalesce around sustainability themes like energy efficiency standards and pollution abatement fees.

Climate science was in its embryonic stage in 1970, even though we were not many years away from having demonstrable human impacts on climate. Only a few dozen papers in climate theory and modeling, plus a handful of meteorology and oceanography texts, were worth reading. Significant developments in the two machines essential for studying Earth's climate and Earth as a system, the satellite and the computer, were occurring at the time I started to pursue systematic studies. Plus, I didn't have to read hundreds of articles, as I did in plasma physics, to get up to speed. I could absorb what was being done in climate modeling by carefully reading 20 pieces or so. Climate science was a budding field of opportunity. Nor was I alone in my excitement. Other scientists and researchers were beginning to suspect the seriousness of what today ranks as one of the most pressing global threats.[5]

One of the most serendipitous events in elevating the threat of adverse climate change in the scientific arena was the hubbub surrounding our paper in *Science,* which occurred as a result of my attending the three-week meeting on the Study of Man's Impact on Climate (SMIC) in Stockholm in July 1971. I was invited to go along by scientists from the National Center for Atmospheric Research (NCAR) to serve as a rapporteur to record notes on the proceedings.

In April I had attended the American Geophysical Union meeting in Washington because there were two talks I really wanted to hear. Carl Sagan was giving a lecture on planetary astronomy and atmospheres, and in another talk, William Kellogg would discuss man's impact on climate. Kellogg had been the lead in the climate chapter of the Study of Critical Environmental Problems (SCEP), which was one of the first assessments of environmental issues.

The aerosol cooling issue back then was mostly defined by Reid Bryson from the University of Wisconsin–Madison, who asserted that biomass burning and desert dust was going to lead to cooling. No one had yet calculated by how much industrial pollution could lead to planetary-scale cooling. Rasool and I did that by converting industrial pollution to sulfate aerosols. Will Kellogg largely discussed greenhouse gases, but also aerosols, mainly because of Bryson. Will was a good speaker and a very personable guy, so I walked up to him afterward and started talking to him. Eventually, I told him what Rasool and I were doing. I gave him a preprint of the *Science* article and showed him what calculations we had made.

Kellogg asked me a few more questions and then asked, "What are you doing in July?"

"July, this July? I don't know."

"We are going to have an entire meeting just on the climate change issue called SMIC, the Study of Man's Impact on Climate, and I need a rapporteur. You seem to me to be exactly the kind of energetic young guy who I would like to have."

He had known me for ten minutes and invited me on the spot to be his rapporteur at a meeting that was going to include such notables as Mikhail Budyko, the director of the Main Geophysical Observatory in Leningrad; Syukuro "Suki" Manabe, who pioneered using computers to model global climate change; and Hermann Flohn, the leading climatologist in Germany. Together they were going to define the field of man's impact on climate—would I go? I would have canceled anything I had planned, including a wedding, to go to that!

"There is only one proviso, Steve," Kellogg said. "You're a rapporteur and therefore you have to defer to the senior guys—just write up what they say."

"Agreed."

Off I went to the SMIC meeting, and it was every bit as exciting as I hoped. Manabe was a dynamo, in my opinion by far the most brilliant and pioneering of the numerical modelers then working on climate change. Budyko, despite the radical simplicity of his model, was a broad climate intellectual. He wasn't just interested in climate change per se, but in how it affected hydrology and paleoclimates. If we can't explain historical ice ages, then how can we trust what we are predicting for the future? Budyko was interested in the human impact on climate and how climate would impact humans. Today his methods would seem trivially simple, but he had no access to state-of-the-art computers. He was doing his equations analytically, so he had to simplify them to the point that they could be solved by a hand calculator. Little did I know that this was the last time he would be let out of the Soviet Union for 15 years.

As it turned out, Giichi Yamamoto from Japan had the most exciting calculation. He had globalized sulfate aerosols in a simple climate model, and he predicted that they were going to lead to an ice age. The not-yet-published paper by Yamamoto and M. Tanaka was being featured in the SMIC report.[6] Here I was at the meeting, with a manuscript submitted to *Science* with Rasool, but I was under gag order—defer to the seniors. The lone contribution I made was my calculation relative to what Manabe and Richard Wetherald showed in 1967—that when you increase cloud amount, you cool the planet. Yet they also showed that the rate of cooling was radically different for low clouds and high clouds, and that very thin, high clouds actually would warm, because their infrared greenhouse effects would dominate their solar reflectivity effects, which increase albedo. (Albedo is the fraction of light that is reflected by objects such as Earth as a whole, or just a part of the surface, or a cloud, or the surface plus an aerosol, and the like.)

I showed that if you didn't change the amount of clouds, but you only increased the height of the tops of the clouds, you actually increased the amount of heating of Earth.

Manabe loved it. He said, "Wow, that's fabulous, you've got to put this in the report!" I think that was the first time the term "cloud feedback" had ever been put into a paper. The book that MIT published from this conference is still, in my view, current in that it set up the basic problems, even though many more-accurate findings have been made since.[7]

Then there was the Australian scientist Sean Twomey, who had a radical new idea—which was featured in the SMIC report 30 years ahead of the IPCC assessment reports—that if you increase the number of cloud condensation nuclei by adding small aerosol particles from sulfate pollution, for example, you will increase the albedo of clouds. Therefore, you could not just look at the direct effect of aerosols in between clouds—there would also be an indirect effect of aerosols on cloud albedo, which Twomey thought would be a radical cooling, more dramatic than what Yamamoto and Tanaka (or Rasool and I) had calculated.

The Rasool-Schneider paper appeared in *Science* four days before the end of the SMIC meeting. Victor Cohn, the reporter for *Science* at the time, called Rasool, who told him that his co-author Steve Schneider was at the Study of Man's Impact on Climate conference in Stockholm presenting our results, which happened to be not true. I was merely there to take notes and write up what the others had said. The day before the press conference in which the meeting would present its conclusions to the international press, I spent the whole day in Stockholm, my first day off in three weeks. When I walked in for dinner, the room started buzzing. Someone intoned, "The iceman cometh." I didn't know what the hell was going on.

Then Kellogg and Manabe came over and said, "Why didn't you tell us?" Yamamoto was sitting in the corner, politely fuming. That Saturday's *International Herald Tribune* had carried a story from the previous day's *Washington Post*—an interview with Rasool about our paper just published in *Science,* saying that I was presenting the results to the Study of Man's Impact on Climate in Stockholm. But I hadn't told anybody about these results. To make matters worse, Yamamoto and Tanaka had just made a similar discovery that was now scooped by the rapporteur, and Yamamoto was the senior professor. After the teasing about ice ages and so forth was over, I gave a little mini-seminar on what Rasool and I were doing. My mentors were really supportive, and even though Yamamoto and Tanaka were featured in the report, Rasool and Schneider would now be included too, because how could SMIC ignore the initiating article from *Science*?

At the press conference the next day, I had a foretaste of the media barrage that was going to follow me through my entire career. I was sitting in the back of the room while the superstar scientists made their presentations. The first question the reporters asked afterward was "Where is Dr. Schneider?" The European press and stringers from the American press started taking notes and sticking microphones in my face. "When is the ice age coming?"

It was media baptism by fire—or rather, ice. I had had a little media experience because of the Columbia student negotiations, but that was talking to the campus paper, not the Swedish national newspaper or the

Times of London. I was smart enough to be careful and say, "No, we did not predict an ice age. What we said was that five degrees cooling, according to Professor Budyko right over there, would trigger an ice age—ask him." I was pretty proud of myself for coming up with that way to deflect the excessive attention. Before long, though, I'd be in the hot seat again.

By the time I returned to New York, Rasool had received many angry invitations, demanding that we explain our "irresponsible" paper. Rasool said, "You did the calculations—you go out and defend it." I went to a number of institutions, terrified that I would fall flat on my face when discussing the work with real meteorologists. I also went to NCAR in Boulder to give a talk, at the invitation of Will Kellogg and Phil Thompson, who helped us write the SMIC report in Stockholm. The talk did not lead to another session of "How dare you publish, Rasool and Schneider?" Instead, they asked, "Why don't you come to NCAR?" They wanted me to become an Advanced Study Program postdoctoral fellow.

A funny chain of events had propelled me to Boulder. Before I gave my talk at NCAR, I had read an opinion piece in the *New York Times* by Bob Guccione, later publisher of *Penthouse* and *Omni* magazines, at that time working for a mining magazine. He wrote a tongue-in-cheek attack on global climate change, saying in effect, "Well, it could be warming from greenhouse gases or it could be cooling from aerosols, but don't worry, folks, if they cancel each other out it's neither." I wrote back a letter in which I said, in effect, "It's real cute, but we don't know whether the warming or the cooling is going to dominate, and the big problem is rapid change from the present, because both agricultural systems and ecological systems are adapted to the present—and we don't want large changes, and this is not anything that should be mocked, because we'd have to be mighty lucky to have the warming and cooling exactly cancel out."

The *New York Times* wrote back to me a day or two before I left, saying they wanted to use my letter—and by the way, where are you? I had sent the letter as a private citizen. I told them I was an atmospheric scientist working at the Goddard Institute for Space Studies but made it clear that I was writing just as myself. "Oh, we only need that for identification purposes," said the *New York Times*.

Somewhat later, I was sitting in Will Kellogg's office at NCAR after giving my talk when he received a call from Bob Jastrow, my boss at GISS. Will began to frown and said into the phone, "Yes, yes, he's here," looking straight at me. Jastrow's next words were, "Good, keep him. I just fired him." Jastrow was steamed because I had published a piece in the *Times* without first clearing it with him.

I called Rasool, who was furious that I had been fired. I found out later that he immediately phoned Morris Tepper, the director of meteorological systems at NASA headquarters, who had actually been encouraging GISS to get into climate studies. Tepper turned around and called Jastrow to congratulate him on my letter in the *New York Times* and to say how really good it was and how much the headquarters approved of hiring "young scientists like Schneider." Jastrow quickly rescinded my firing, but when the opportunity at NCAR came through several months later, I didn't waste any time accepting it.

During my NCAR years, the fundamental mission of the institution would come under radical reexamination. The emerging technology of the computer age would permit research projects on a scale never experienced. We had to ask fundamental questions. Were the results of those climate and atmospheric models "real" science? Was there value and integrity in expanding our work beyond science to societal problems? If the indications that humans were having an unforeseen and rapid effect on the environment were accurate, what was our responsibility now and in the future? What happened during the next decade profoundly changed the way climate science is conducted.

DRAWING UP THE BATTLE LINES

2 **CLIMATE SCIENCE WAS** gradually expanding. My years at NCAR in Boulder would be marked by increasing sophistication in predictive models, driven largely by improvements in computing technology. Yet the very idea of creating models met stiff resistance from some empirical science types, who claimed that only hard scientific facts—actual observations—could be credible. But how could one apply the traditional scientific method, the empirical approach of observation, experience, or experiment, to the future?

I argued for an expansion in our thinking: A hierarchy of approaches needed to be considered. Planetary data were certainly necessary but not sufficient to make credible predictions, and we had to be able to draw reasonable conclusions about the future, not just from what we could measure, but also from what we could model. After all, there can be no data for the future before the fact, so any prognostication into that unknown territory was, by definition, a model of the factors that determined how the future would evolve.

The pioneers in Earth systems science would also encounter opposition as we tried to define our discoveries in the broader context of what effects pollution was having on society as a whole. The work we were doing would lead to a multidisciplinary approach, in which experts from a variety of fields, such as agriculture, meteorology, chemistry, ecology, sociology, and so on would study the problems of climate change and draw conclusions that could affect all humankind.

The excitement of that time was starting to build even before I left Manhattan. Jim Hansen and I used to grab lunch at Tad's Steakhouse most days when I was working at NASA's Goddard Institute for Space Studies (GISS). Over a five-dollar cafeteria steak and baked potato, a bunch of young scientists would argue about the future of our research.

"Why are you working on this climate stuff?" Jim once asked, possibly tongue in cheek. "It's such an impossible problem. It has so many dimensions; you can never fully solve it. Do something tractable like radiative transfer in planetary atmospheres and clouds."

Jim, even if kidding, was right, of course.[1]

We can never fully solve the climate prediction problem. But we can go a long way toward bracketing probable outcomes, and even defining possible outliers. Jim and his colleagues at NASA have done some of the best work anywhere trying to quantify outcomes. As director of GISS since 1981, he's been fighting a broad battle for action to limit the effects of anthropogenic climate change and has been involved in more than his share of controversies. His testimonies before Congress during the 1980s were drawn from his innovative work on aerosols and other atmospheric gases, which he expanded from his graduate school studies of the atmosphere of Venus and radiative transfer (the physical transfer of energy in the form of electromagnetic radiation, waves of varying frequencies such as infrared or visible light).

A physicist and astronomer, Jim applied the mathematical equations and computer codes that he developed for his Venusian work to the study of Earth's atmosphere. He was exploring the possibility that Venus, the hottest planet in the solar system, had had an early atmosphere similar to Earth's and that increased warming due to the inexorable radiant energy trend of the sun, combined with Venusian greenhouse gases, had led to its present atmosphere—composed of vast quantities of CO_2 and sulfuric acid clouds—and its very high surface temperature.

Jim's research conclusions evolved as the capabilities of scientific research became more sophisticated and GISS developed tools to study Earth's climate in more detail. But in the beginning, the idea of using general circulation models (GCMs) to examine Earth's climate was radical. They were initially developed primarily as a weather forecasting tool, and no one really knew how credible they would be for climate change projections. At GISS I worked on very simple models of the energy balance of Earth—that is, the difference between the amount of sunlight absorbed by Earth compared to the amount of planetary heat radiation back to space—to see if climate was stable when the normal energy flows are disturbed by factors such as greenhouse gases and aerosols. The climate is relatively stable, up to a point, we found. I also investigated how clouds serve as a feedback mechanism. Clouds either amplify climate change in the direction it was already going (so-called positive feedback—"positive" here *not* meaning good for you), or they resist change and thus stabilize the climate system as it was being disturbed (known as negative feedback, "negative" here *not* meaning bad for you).

At the same time I conducted a joint project with a young meteorologist at NCAR, Warren Washington, who, with Akira Kasahara, had developed a general circulation model competitive to that at Princeton's Geophysical Fluid Dynamics Laboratory (GFDL) and GISS. I never regarded the collaboration as disloyal, but I'm not sure everyone at GISS was happy about it. Warren and I used NCAR's general circulation model to study cloudiness as a climatic feedback mechanism. Our insights were elementary and pretty naive relative to today, but helping to define the big questions during the early days of climate modeling was very exciting.

At GISS, another of our young companions, Richard Somerville, was very productive in working on improvements to the general circulation model that GISS had adopted from two UCLA professors, Yale Mintz and Akio Arakawa. Richard was a theoretical meteorologist who had been working a few years at GFDL in Princeton with Syukuro Manabe, the best climate modeler in the world then. Richard became an expert on computer simulations of the atmosphere, and he would be a coordinating lead author for the IPCC 2007 Fourth Assessment Report. I have been lucky to work shoulder-to-shoulder with such distinguished scientists over pretty much the whole history of climate change research. When I moved to NCAR in 1972, I was to meet many more.[2] During those years at NCAR we would help to build a new field that would blossom into the scientific underpinnings of the current climate change debate.

NEW IDEAS FROM GENERAL CIRCULATION MODELS

My first experiences in Boulder would be circumscribed by how much we didn't know. Warren Washington was excited about what I was doing on cloud feedback, and together we evolved a series of general circulation model experiments where I increased and decreased the fixed ocean temperatures in the model. In those days, we did not calculate ocean temperatures but assumed—because of computer limitations—they would stay the same over time and area.

Using NCAR's GCM, I increased the ocean temperatures by two degrees Celsius and decreased them by two degrees, and we calculated what happened to the cloud cover. We discovered the correlation was a positive feedback—as the ocean warmed, cloud cover decreased—suggesting an amplifying effect, and the reverse occurred for decreased ocean temperatures. That conclusion flew in the face of conventional thinking, which held that clouds created negative feedback—for example, as a summer day warms up, the likelihood of thunderstorms increases and they stabilize the system by cooling it back down.

I subsequently conducted two strip experiments. I wanted to see if there was a difference between warming up the whole ocean and warming up only part of it. I warmed up a strip of ocean between the Equator and ten degrees south latitude, and nowhere else, and studied the cloud responses of the model. It resulted in very little local effect but caused clouds to decrease significantly—positive feedback— over the subtropics of the Northern Hemisphere because it strengthened the north-south atmospheric circulation. We call long-distance responses teleconnections, and they are common in climate dynamics. A well-known example is the warming of the equatorial eastern Pacific—an El Niño event—that has teleconnected repercussions half a world away, such as floods in California.

I also warmed a strip of ocean between ten and twenty degrees of latitude in the Northern Hemisphere and garnered a weird response that altered the general circulation of the atmosphere, adding clouds above the strip but decreasing them in adjacent latitudes. What we found from this kind of work is that a great deal can be learned using fairly simple models of the climate system that account for basic atmospheric dynamics and for radiant energy going back and forth between Earth and space, despite the absence of an ocean that could change its temperature. Yet if we were to predict regional climate effects from adding greenhouse gases or aerosols or modifying the land surfaces, we would need a three-dimensional representation of our virtual reality computer model that included a more realistic ocean and would thus have to model in spatial detail what might happen.

In 1973, Bob Dickinson and I, building from the SMIC report, classified this need for a "hierarchy of climate models," from the simple Earth-averaged ones (which just calculated radiant energy but were transparent in identifying cause-and-effect mechanisms) to the very complicated atmospheric-oceanic general circulation models (which included representations of all known important processes, like ocean dynamics, cloud feedbacks, ecological changes, and ice and snow changes).

Today the scientific community primarily uses GCMs that include all of these processes—but debate persists on how much detail is really needed to make credible projections. Assessments such as the Intergovernmental Panel on Climate Change (IPCC) reports help to clarify the state of the science every six years, but in truth it is a moving target, because the more we learn, the more we find that we need to go back and change some components in the models; we then try to replicate our findings to build confidence in any inferences.

It takes a community of scholars from different institutions and countries, repeating with different computer models what others have done, to form intercomparisons and to pioneer new models. Back in the early 1970s, when a reporter asked how long this process of model improvement, replication, and predictions would take to achieve high confidence, I recall saying that our models were "like dirty crystal balls, but the tough choice is how long we clean the glass before we act on what we can make out inside." That risk management challenge plagues us still, even as models become more sophisticated and simulate Earth's conditions increasingly well. What constitutes "enough" credibility to act is not science but a value judgment on how to gauge risks and weigh costs.

The ocean surface warmings I imposed in Warren's model were for a "perpetual January" model, meaning that the sun was held fixed in a January position, and observed January sea surface temperatures were also fixed. I wasn't happy with that condition because it meant the model was not conserving energy. I couldn't achieve a balance between the incoming solar energy absorbed and the outgoing infrared radiant energy, which happens when the temperature changes enough to restore balance. When temperatures in the ocean are fixed, it becomes an infinite sink or source to absorb or give up the heat, respectively. We now know that the effectiveness of the world's oceans as a heat sink is limited, not infinite, and that global temperature is inextricably bound to ocean dynamics and thermodynamics. But that was too difficult to calculate then.

CREATING CLIMATE-ORIENTED SCIENCE FROM SCRATCH

Warren and I wrote a joint paper on our GCM work, which I presented in Sendai, Japan, at the International Radiation Symposium meeting in the spring of 1972. I had the fun of meeting people like Tom Vonder Haar, a Colorado State University atmospheric scientist who was working with satellites. A half dozen of us went down to Kyoto on the fast train, and I remember sitting in Japanese restaurants teaching some of my new friends how to use chopsticks. Having lived in New York, I was already acculturated to Asian food. We were pretty much the only Westerners in the restaurants. In one place, six-foot-six Vonder Haar—a big, big guy, not a standard stature in Japan—was trying to fold his legs underneath one of those Japanese tables where you sit on the floor, and the waitresses were looking at each other and trying to be polite and not laugh. We sat there for about five minutes and then all of a sudden, like an undone spring, his legs uncoiled—the table went bouncing, the sake went flying. The waitresses covered their mouths and giggled hysterically. Tom took it very well.

I started to develop a sense of community within the radiation wing of the meteorological community. A global organization like IPCC, combining the research and ideas of a thousand scientists and policymakers, was not on anyone's agenda back then. When I started at NCAR in 1972, the combining of multidisciplinary efforts in scientific study of climate was a revolutionary concept. But that's what we hoped to do at NCAR. We had vision and enthusiasm; we set out to do it.

Will Kellogg ran the Laboratory of Atmospheric Sciences, which actually included chemistry, physics, and dynamics as three separate groups. In November 1972, after I was at NCAR for all of three months, he decided to run a retreat for the staff in Winter Park, Colorado, which already had heavy snow. I still remember trudging around happily in cross-country skis up there, with fellow Columbia plasma physicist Bob Chervin, now working with me at NCAR, thumping around in snowshoes, walking up a hill at 9,000 feet with much huffing and puffing.

One night we were all gathered, drinking plenty of wine, and I had been asked by Bob Dickinson to give a ten-minute talk about why NCAR should undertake a serious expansion in climate research. I didn't discuss climate from the point of view of GCMs alone. First, I posited the idea of a hierarchy of models. That was a concept that Suki Manabe, Mikhail Budyko, Phil Thompson from NCAR, and I had come up with in the SMIC report, and two years later Bob and I refined the concept in-depth in a survey article that we wrote in *Reviews of Geophysics and Space Physics*. We tried to define the field of climate modeling, which had never had a survey before and had no single accepted set of definitions about where it should go.

I argued that with a hierarchical approach, we don't need to understand a problem in all of its depth. We can run a hierarchy of models with the simple ones addressing basic questions, and large-scale three-dimensional models to simulate some of the complexities of reality. I had already learned from the work with Warren Washington that all kinds of interesting displacement behavior and teleconnections could be discovered when you run GCMs.

The simple models are often called, somewhat playfully and pejoratively, "back of the envelope models" because they may be only one equation long—which could be scribbled on the back of an envelope, and not take a page or two of advanced math like a GCM. Everybody was a little bit inebriated by then, and I was using an opaque projector to demonstrate a simple energy balance equation—naturally, I scribbled it on the back of an airmail envelope to enhance the gag. Some people were laughing and playing along, but some people, I noticed, were humorless and hostile. Computer models back then were regarded with suspicion by traditional empirical scientists.

Walter Orr Roberts, founding director of NCAR and president of the University Corporation for Atmospheric Research (UCAR), spoke at the retreat as well, and he announced that we were going to experience a radical reconsideration of the institution. I hadn't had much chance to meet Walt, but I admired him. He was interested in the societal applications of science. He had founded NCAR in 1960 originally as the High Altitude Observatory, and then he broadened its scope to include atmospheric science. Now he wanted to move in the direction of climate, because he saw that as an evolving social and scientific problem. He also had to satisfy the 50-odd universities of the UCAR consortium, which apparently had complained about NCAR moving too much toward big science and away from basic science. At the same time, they asked NCAR not to compete with them directly for limited federal research funding.

In response, when we returned to NCAR, Bob Dickinson, Will Kellogg, Phil Thompson, and I proposed a climate project. We worked long and hard on the proposal. We suggested the effort be based on the hierarchical approach: We would work on simple models; work across the hierarchy; connect with the observationalists, for validation and derivation of so-called parameterizations (attempts to incorporate processes not explicitly modeled but that had to be included by approximations); and make cloud feedback a central component. The proposal wasn't very long, maybe 10-15 pages. Will's long-term administrative assistant, Barb Hill, typed up the proposal and handed it to Will to proofread. He just started laughing.

I said, "What's so funny?" I looked at the opening line, which began, "We propose a conceited effort at climate research at NCAR. . ."

"No, Barb, that's *concerted*."

She looked at me. "I was just thinking about Steve, you know."

In the winter of 1973, the project rankings came back, and our climate project proposal was number one. I was then asked to be the deputy director of the project, working in a real job with Phil Thompson. While the offer was heady and flattering, at the same time it was intimidating—I had been an NCAR postdoc for only six months. It also earned me a host of enemies, who were furious that, because of budget cuts, we were displacing "real" scientists, who make measurements, in order to do this theoretical modeling stuff.

SPOTLIGHT ON CLIMATE STRATEGY

Any expansion of the climate project would benefit by including others besides our group at NCAR, so I walked into Will Kellogg's office one day and said, "We need to demonstrate that there's a genuine interest in this." Instead of having the usual stuffy scientific seminars, I proposed we have a Climate Club. The club would send out an abstract of a talk, and a title, and we would try to get interesting speakers to take on controversial subjects. We would, for instance, bring in J. Murray Mitchell, a pioneer in the analysis of climate data to show climatic trends, to discuss the big issues—is Earth going to warm, or is it going to cool? Bring in Bill Sellers from the University of Arizona, whose controversial model led Rasool and me to propose that global cooling might lead to an ice age.

Will responded, "Well, if we want to make it interesting, and we want to make it a club, why don't we have the talks late in the day so that when they finish at five o'clock, we'll have wine and sherry and cheese?" Will was very clever.

Between the two of us, we worked out sexy topics and a bribe of grape ethanol to get everybody in the room, and the series worked like a charm. We had 80 to 100 people attending these meetings. They were fun. I would introduce the speakers, always trying to be humorous, and everybody had a good time. The ones who hated our style simply stayed away. Tellingly, the Climate Club often outdrew the regular seminars significantly, which indicated to the NCAR management that climate change research was a topic that had a growing constituency.

I wanted a speaker from a new global think tank called the Club of Rome. Their first published report was a best seller at the time, warning that unlimited consumption and population and economic growth on a planet with limited resources cannot go on forever and is dangerous. Opponents called them neo-Malthusians. I got John and Cheri Holdren to come out in 1974, and John did a more credible job of laying out the population-resources-environmental problems than nearly anyone else could have done at the time. That talk helped the NCAR scientists to see the big picture clearly and early on.

I did get an even more famous neo-Malthusian, though, Stanford University's Paul Ehrlich, author of the 1968 best seller *The Population Bomb*, to come talk about the role climate played in the so-called world predicament. Paul was a regular guest on Johnny Carson's *Tonight Show* as a science and policy expert because he was both very funny and very insulting to those he saw as mucking up the planet. He not only filled the NCAR seminar room, they had to put speakers outside. I told Paul that his job was to convince my somewhat reluctant atmospheric science–oriented colleagues that for us to move forward into the world of real climate applications, we needed to do more than just study meteorology or oceanography. We needed to involve ecologists, economists, and others in a broad multidisciplinary program.

Paul was trying to convince people of just that, until someone from the audience stood up and said, "Dr. Ehrlich, all of what you say may be interesting, but not everybody agrees. There are even members of the National Academy of Sciences who do not share your pessimistic views" on global famine and overpopulation. Someone else chimed in to say that perhaps we should allow our elected officials to solve these difficult problems for us.

Paul, always the in-your-face humorist, uttered one of his infamous statements: "You know," he said, referring to these exalted academics, "it isn't only cream that rises rapidly to the top—and floats." It brought the house down—well, most of it—while enraging others. By that point it was obvious that we were going to have to live with a mixed bag of people, some who wanted to turn NCAR in the direction of becoming a global climate change institute—and some who genuinely thought this broader effort was anathema to the scientific purpose of the institution.

The opposers' basic view was that if you weren't making field observations, doing lab experiments on a narrowly focused set of questions, or writing down and solving three-dimensional equations, you weren't doing real science. In my opinion, if you weren't trying to understand what mechanisms were causing these models to give you the answers to key problems of planetary relevance, you weren't doing important science. We needed to do both. We needed to explore with stripped down, highly parameterized models, and then ask the same questions across the hierarchy and see what we could learn.

In the fall of 1972, accompanied by Bob Dickinson, I went to Fort Lauderdale, Florida, to the Scientific Panel on the Natural Stratosphere, the very first Climatic Impact Assessment Program (CIAP)—an early precursor to the IPCC. The Boeing Corporation had proposed that a supersonic transport (SST) be federally subsidized, to compete with the European's Concorde jet, and many environmentalists were upset because it had been reported that the injection of water vapor in the stratosphere would cause ozone depletion. (It turned out later that the problem was nitrogen oxides, not water vapor.) We met a young postdoc from the Lawrence Livermore National Laboratory, Mike

MacCracken, as well as Richard Lindzen, then a young atmospheric physicist at the University of Chicago. I would serve on the climate panel with MacCracken, Dickinson, and several others.

On day one, Lindzen denounced the enterprise as irresponsible. The report had to be written in two years, and he claimed that the science would not be definitive in that time frame. According to him, whenever scientists are forced under political pressure to provide answers that cannot be given, they are violating their scientific integrity. Although Lindzen had very impressive credentials, I totally disagreed with his scientific philosophy, as did Mike MacCracken and Bob Dickinson.

MacCracken said, as I recall, "Dick, we are not arguing that we should tell people we know the answer. What we are basically saying is that we have the best information there is and that we should explain what we know, and what the likelihoods are, and how much of this we can fathom. And we should say what research needs to be done." Mike outlined beautifully what a good assessment should do. I thought it was an excellent speech—and in essence presented the same arguments for risk assessment of climate change that I hammer home in my talks today.

Lindzen excoriated him. Basically, he said, "You don't understand. That's not what science is. These politicians should never push us around, and we should not give in! We should demand that we scientists are in charge of the scientific agenda. We are a better enterprise than politics."

I stood up at that point. "Dick, they are going to vote on the SST in a few years. They need the best science that's available. Do you think it's more responsible for us to guess with caveats attached, or to have Senator Barry Goldwater guess for us?" A couple of people applauded.

Lindzen turned around and said, "That is the most scientifically irresponsible thing I have ever heard." He stormed out of the meeting shortly thereafter. Many of those in attendance agreed that we were trying to do an honest job of assessment. This group was so far ahead of the times that it included economist Ralph D'Arge from the University of Wyoming. They actually wanted him to assess the economic implications of ozone depletion. Of course, nobody else in the room knew anything about economics, and he knew nothing about climate. Now when we conduct global climate change assessment meetings, many people have a pretty good understanding of the wide range of topics.

As part of the assessment report, I would help write the chapter on climate along with Chuck Leith, a senior atmospheric physicist at NCAR, and Jim Coakley, an atmospheric physicist and postdoc at NCAR, whom we hired as a project scientist. As a result, I had an opportunity to recalculate the CO_2 warming versus aerosol cooling problem from my GISS days, when Rasool and I predicted that if CO_2 concentrations in the atmosphere were to double, the planetary surface temperature would warm up a mere 0.7 degree Celsius (1.3 degrees Fahrenheit). Suki Manabe had done a series of calculations in the 1960s that suggested this number should be at least twice that. Nobody knew what the answer was, but it was terribly important to determine how much the system would warm up—what has been called "climate sensitivity." It is the most important single prediction from climate models—then and now.

Jim built a radiative-convective model like the one Manabe had a decade earlier. Because we had to consider the stratosphere for the SST problem, Jim's model included the region above 16 kilometers (10 miles) high, unlike the model that I had adopted for the Rasool-Schneider paper. That's how I discovered what was wrong with my CO_2 calculation, because when we included the stratosphere, instead of 0.7 degree Celsius for climate sensitivity, the new results showed 1.5 degrees Celsius (2.7 degrees Fahrenheit). Rasool and I had missed the downward infrared radiation enhancement that occurs when you add CO_2 to the stratosphere. That factor turned out to be, as shown by Ben Santer at Lawrence Livermore National Laboratory some two decades later, a key to IPCC attributing the warming of the planet's surface and cooling of the stratosphere to ozone-depleting chemicals and greenhouse gases from human emissions.

Around the same time, Kellogg was beginning to work with Gerald W. "Jerry" Grams in the chemistry division of NCAR on what he playfully called GNP, or gross national pollution, where they were actually drawing maps of regionally patchy aerosol loadings that were concentrated around industrial areas, not globally widespread. The excitement of that advance might be hard to imagine now, but we still had so much to learn about identifying global greenhouse gas pollution, aerosol hazes, and the ozone problem.

By 1973 I was convinced that the Rasool-Schneider calculation couldn't be right, because we now had found good measurements of the geographic distribution of aerosols, and they were not uniformly global in extent, but they were regionally significant. I immediately proceeded to read all the papers that had been done that estimated climate sensitivity, and then plotted them all out on the same figure. In my opinion then the best guess of climate sensitivity was between 1.5 and 3.5 degrees Celsius (2.7 to 6.3 degrees Fahrenheit), based upon the literature, which I published in 1975 in the *Journal of Atmospheric Sciences* in a paper called "On the Carbon Dioxide Climate Confusion."

I am very proud of the fact that I operated in the best tradition of science: You draw conclusions based on what you think at the time, making all your assumptions explicit; then you reexamine the assumptions in light of new evidence; you recalculate; and then you publish the revisions without any shame. That's how science proceeds. A model provides the logical consequences of explicit assumptions. The real science is in how good the assumptions are—and that is where empirical testing and peer debate comes in.

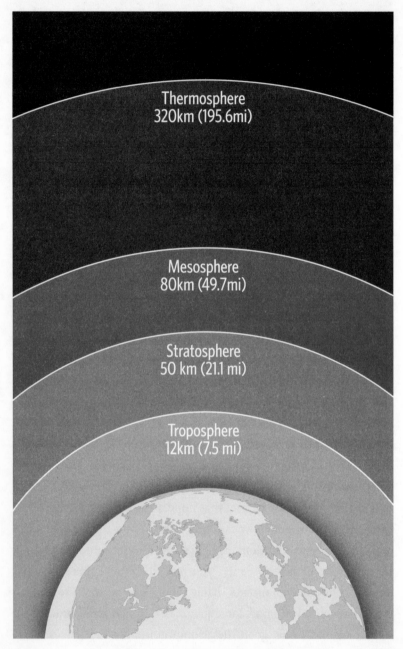

The structure of Earth's atmosphere

As I mentioned earlier, I was attacked decades later by George Will, Charles Krauthammer, and others about predicting cooling in the early 1970s as a grad student and now warming as a senior scientist. In a rebuttal in the *Washington Post*, I wrote, "Imagine the doctor who makes a preliminary diagnosis before the blood test and the x-rays are in, and then they are different from the preliminary diagnosis, but the doctor sticks with it to be politically consistent. This is not what we do in science . . . and we're not ever ashamed of getting the wrong answer for the right reason."[3]

In 1973 Bob Dickinson and I wrote a survey paper on climate modeling for the *Reviews of Geophysics and Space Physics*. Bob had reviewed my initial draft and returned it with so many red pencil marks on it that I decided to ask him to co-author it with me. We complemented each other well. I was writing a paper that would be accessible to a wide readership across many disciplines, and he was trying to make it as technically thorough and well grounded as possible.

I also sent a copy of the paper for review to Mikhail Budyko in Leningrad, and I became concerned when I didn't hear back from him. We had maintained correspondence, and I had sent him other articles and received good comments from him. By late 1973, though, Budyko was becoming persona non grata in the Soviet Union, apparently, I was told, because he had a number of Jewish scientists in his institute. When he was ordered to fire or demote them, he refused. In retaliation, he was heavily censored, was not allowed to go in the Western literature room, and was banned from travel.

As the galley proofs were coming back on the climate modeling paper, Chuck Leith came up to me and said, "I was in Russia last week, and I'm playing mailman." He handed me a crumpled sheaf of yellow lined paper from Budyko, a handwritten review and comments on our paper. Chuck said sadly, "He gave it to me in the bathroom, where he thought we weren't being watched." That was the Cold War.

I incorporated a number of Budyko's comments in the galley proofs, and I know that he was pleased that he was quoted and thanked in the acknowledgments.

Bob and I earned an honorable mention in the NCAR publication prize competition for the paper, which was very well received internationally. Climate science was indeed shifting into second gear. Joe Smagorinsky, who didn't believe in simple models, grudgingly said, "Well, this is a fine didactic work, but all the real modeling is going to be done in 3-D." And I said, "Yes, Joe, we have to use 3-D models increasingly as computer power grows, but we still have to understand what we are doing, and the best way to do that is by using the hierarchy of models approach."

I still believe that principle to this day. You don't just add resolution and complexity for its own sake. You still have to disaggregate into models of varying complexity to see how different subsystems and processes interact, and shortly thereafter you simulate it in three dimensions. Before simulating in full-blown three-dimensional glory, we first have to understand the behaviors of each subsystem separately, and then see what happens when the subsystems are linked. I later called this linking process the "simple simulation of complex models." Working with Starley Thompson, a graduate student from Texas A&M University, and Dave Pollard at NCAR, we created models of glaciers, oceans, and atmosphere. Each of those models was very low resolution (one dimension at best), but we would use exactly the same coupling physics that the three-dimensional ones used. Those simple simulations of complex models enabled us to explore the coupled behavior within our computer budget.

Often when you couple models together the answers are different from what could have been found by running each submodel individually. These are known as "emergent properties"—that is, behaviors that are synergistic and emerge from interactions among several subsystems. I sometimes worry that we are overeducating our students on the detailed analysis of the most complex models, while undereducating them on the need to look for emergent properties using a hierarchy of simpler tools.

FORECASTING PROBABILITIES, MEASURING THE FUTURE

This question of predicting the future has been so thorny that I feel compelled to explain why it caused such a rift. Shortly after joining NCAR I was invited by Robert M. White, the first administrator of the National Oceanic and Atmospheric Administration (NOAA), to give a talk on human impact on climate at the American Association for the Advancement of Science (AAAS) in Baltimore. Helmut Landsburg, whose well-known textbook on physical climatology was published in 1942, served on that panel, and he told me bluntly that modeling was "no damn good"—you just measure nature and learn from that. Of course, I agreed, we must work from empirical evidence but, I argued, you can't measure the future before the fact. If you don't model, you don't know anything about the future. Measuring is trying to understand what happens, but to forecast you need to model processes.

What many empirical scientists missed—and some still do—is that once someone associates two sets of observed data and thinks it can be used to forecast, then an implicit model has been constructed. The old saw "correlation is not necessarily causation" is alive and well. The models constructed from correlations are important steps in formulating hypotheses, but the hypotheses need to be tested by a wide variety of techniques. Data doesn't predict by itself, it only suggests processes at work that must be explicitly modeled. Projections will be as good as the models, built on as much relevant data as possible.

That was a very controversial view then. "Science is empirical," I was repeatedly told. But science is not empirical when you're discussing the future. There is nothing empirical about the future. At the time, I didn't understand Bayesian versus frequentist statistics, but in fact that was about the heart of the matter—objectivity and subjectivity in modeling and in projections.

An English clergyman and mathematician, Thomas Bayes (circa 1702–1761), formulated a kind of probability statistics now called Bayesian inference; his theorem was published posthumously in 1764. In essence, our knowledge base—and prejudices—establish an a priori probability (that is, a prior belief in what will happen). The more we study the system, obtaining more data and better theories, we update our prior belief—this is called Bayesian updating—and form a new belief, known as an a posteriori probability—after the fact. Over time, we keep revising our prior assumptions until eventually the facts converge on the real probability of how various experiments will turn out.

Since we cannot perform experiments on the future before it happens, predicting the future is wholly a Bayesian exercise. This is precisely why IPCC redoes its assessments every six years or so, since new data and improved theory allow us to update our prior assumptions and increase our confidence in the conclusions. Although the scientists in the assessment have reached a consensus, it does not mean they all agree on some conclusion. Instead of complete agreement, what society needs from scientists is to know the full range of potential outcomes as well as how confident we are in those conclusions.

So consensus about future events includes the degree of confidence we have in our projections and conclusions, not simply the conclusions themselves. Some conclusions enjoy a strong consensus, such as the very high probability that Earth has warmed since 1850. Other conclusions, for instance, the possibility of many meters of sea level rise in centuries, have less confidence—maybe only a 50/50 chance. The conclusions cannot be objective, since the future is yet to come. However, we can use current measurements of ice sheet melting. We can compare it with 125,000 years ago, when the planet was a degree or two Celsius warmer than now and sea levels were four to six meters

(13 to 20 feet) higher. Because that ancient natural warming had a different cause from anthropogenic greenhouse gas increases now, we can't say with high confidence that a few degrees warming from greenhouse gases will also cause a four-to-six-meter rise in sea levels. But it undoubtedly indicates an uncomfortable Bayesian probability that we face meters of sea level rise in centuries. This indeed was the conclusion of IPCC's Fourth Assessment Report (Working Group II and the Synthesis Report volumes) in 2007, for exactly those reasons.

Some statisticians and scientists are leery of Bayesian methods, preferring to stick only with empirically determined data and well-validated models. But what do you do when you don't have such data? One example is found in clinical trials in cancer treatments. The "gold standard" in clinical trials is a double-blind trial where half the patients receive a placebo and the other half receive the drug being tested, and neither the patients nor the researchers know who got what. After the test period—five or ten years perhaps—if there is a statistically significant difference between the recovery statistics of patients given the drug relative to the placebo, the trial is declared successful. The trial isn't designed to pinpoint individual differences, only to give aggregated frequency statistics. Even if we knew the odds of recovery for the average person from different treatments, there is a wide spread in individual responses. So best practice medicine should try to individualize treatments to match the individual's idiosyncrasies. That makes some doctors—and many insurance companies—nervous. Likewise, some scientists and many policymakers are nervous about Bayesian inferences based on the best assessment of experts, preferring hard statistics. But as there are no hard statistics on the future before the fact, Bayesian methods are all we have and certainly are better than no assessment at all and taking potluck that all will work out fine with no treatment.[4]

If we care about the future, we have to learn to engage with subjective analyses and updating—as there is no alternative other than to wait for Laboratory Earth to perform the experiment for us, with all living things on the planet along for the ride. In my 1997 book, *Laboratory Earth,* the subtitle was simply *The Planetary Experiment We Can't Afford to Lose.*

That 1973 AAAS meeting in Baltimore added fuel to the conflict. I gave a talk about warming versus cooling, and by that time, I had realized that aerosols were not globally distributed. I could no longer make the claim, as Rasool and I had, that the cooling dominated the warming. I said that, depending upon what assumptions we make, we could get substantial cooling or substantial warming, and that it wasn't yet clear which might dominate. We didn't know enough about the distribution of aerosols. I quipped, "Mark Twain had it backwards. Nowadays everybody is doing something about the weather, but nobody is talking about it."

A white-haired gentleman sitting in the front row was writing all of this down on a small reporter's pad. That gentleman was Walter Sullivan, the dean of science writers at the *New York Times*. He quoted the crack about Mark Twain and then wrote about how the climate could warm or cool from human activities. The article was hardly front-page news, but NCAR at the time had a press clipping service, and any clippings mentioning NCAR or UCAR were photocopied and distributed around the institution.

When I came back from the AAAS meeting, I found the *New York Times* story posted on the bulletin board of the NCAR map room, where everybody congregates, with a big blue "BULLSHIT" rubber-stamped on it. Of course, whoever was responsible didn't have the guts to say it to my face. But I then began hearing about people who were furious that "this Schneider guy" was always seeking publicity. Yet I hadn't been seeking publicity at AAAS—it was a quip that came to me at the end of the talk and simply summarized my message in a sound bite that caught the attention of Walter Sullivan. He interviewed me afterward, we talked about these issues, and he did a good job of honestly reporting the session.

In science, people derive a reputation from their scientific publications. But now we were entering a media age, and I was getting a lot of notoriety. Some other scientists were hostile because I was becoming known not on my science alone, but because I could deliver a sound bite that caught people's attention. Unfortunately for me and many others who are effective with public communication and the media, not all reporters are like Walter Sullivan and get complex issues right. The worst thing that can happen to someone in science who plays in the media game is to get a bad story that just gives legitimacy to those who object on personal or philosophical grounds to science popularizers.

SCIENCE FOR POLICY

We were beginning to understand which questions to ask about greenhouse gases, aerosols, warming versus cooling, ocean temperatures, global air currents, and other physical aspects of the planet's climate. We wouldn't have many confident answers for years to come, but we needed to take another step forward as we began to look more broadly at climate science—as a significant research field on its own, beyond just projecting climate statistics.

One dimension we needed was the politics of environmental issues. Around 1974 Walt Roberts added a political scientist to NCAR, Michael (Mickey) Glantz. Mickey had the scientific training to be able to talk with many folks who had an ingrained bias against political issues intruding into science.

At NCAR we really needed a social science perspective. I was, almost by default, doing some social science on the side, by asking broad policy questions about whether research we were doing was relevant to what people in our society needed. I talked to experts in agronomy, for example, and found out that they cared much less about whether climate changed permanently by two degrees Celsius. What threatened crops was the year-to-year variability in droughts, floods, hail, heat waves, and cold spells. Of course, a change in the mean temperature can create changes in variability and extremes.

Mickey knew how to address social science questions. He could ask penetrating questions about whether certain technologies were desirable, and who the winners and losers might be, and who was paying, and who the beneficiaries were, and whether it was appropriate to be messing around with nature or setting various policies. Those are tough questions, and people trained only in atmospheric sciences, in most cases, wouldn't ask them. Mickey would become a well-known champion of helping human beings adapt to climate change. He was the founding director of NCAR's Center for Capacity Building, a small group of experts dedicated to assisting communities in Asia, Africa, and other areas less fortunate than the United States in dealing with the societal impacts of weather and climate.

Asking questions about social impact issues coincided with the redirection of an existing international meteorological program called the Global Atmospheric Research Program (GARP), sponsored by the International Council of Scientific Unions (now the International Council for Science) and the United Nations. GARP's original objective was to help the international community improve long-range weather forecasts. GARP scientists also fashioned a second GARP objective—climate change research. This two-objectives framing forced the scientific community to clearly make the distinction between the forecasting of weather and the forecasting of climate. Climate has playfully been called "what we expect"—and weather is "what we get." More precisely, weather is the instantaneous state of the atmosphere. A weather forecast starts with today's weather—the so-called initial condition—and predicts the evolution of the details of the atmosphere forward in time—the forecast. Climate is the long-term average of weather—typically over decades or more. That is a so-called boundary value problem. Since a change in the energy output from the sun, for example, impinges on the top boundary of the atmosphere, the atmosphere reacts to this forcing, not to the initial state of the atmosphere. Thus weather and climate forecasting are fundamentally different problems, even when using the same tools—like GCMs. How the tools are used is what matters in this difference.

The first objective of GARP was to improve the quality and reach of long-range weather forecasts, by which they meant six to ten days. That effort required a globally comprehensive, more accurate data set of the initial condition. The problem was that we didn't have the richness of data for many ocean regions, especially in the Southern Hemisphere, and in some developing countries that we had for the wealthier countries in Europe and North America.

If you want to do weather forecasts, you have to know how an atmospheric event began in order to run it forward. Many people believed that if a better data set could be used, farmers could be given six-to-ten-days' notice of when to expect weather that was potentially damaging or good for harvesting or planting.

GARP had a second objective, which wasn't pursued as much in the beginning, until it became clearer that the first objective was not likely to work very well. Improving accuracy in six-to-ten-day forecasts was going to be tough. Because credible weather forecasts appeared intractable after about a week, how, critics asked, could we make a credible long-range climate projection over years—well after chaotic dynamics of the atmosphere scrambled any memory of the initial conditions?

I asked meteorologist and mathematician Ed Lorenz—a passionate mountaineer—this question during a hike up 8,500-foot-high Bear Peak, right behind NCAR, in the summer of 1973. "How do you respond to those who ask how we can make a long-range climate forecast if we can't even make a credible weather forecast past a week or two?" He came up with this brilliant analogy. Imagine a pinball machine where we wanted to predict the precise sequence of ball-pin encounters—a weather prediction metaphor—by measuring the initial condition of the ball by a machine that controlled to high precision the initial speed of the ball. Knowing the geometry of the pinball machine and the laws of mechanics, it would be easy to predict the first ball-pin encounter, and the second, too, as the angle of reflection from the pin was equal to the angle of incidence of the ball. And so it goes to the third and fourth pins—or so it would seem—if not for the intrusion of chaos.

As Ed put it, a tiny vibration in the room or speck of dust on the path of the ball that we couldn't predict could make the ball deviate almost imperceptibly from the precisely predicted path. That infinitesimal error would be multiplied every time it hit the next pin until after only a few ball-pin encounters, it would miss the predicted next target altogether—the pin encounter "weather forecast" would reach its "predictability limit," as Ed called it. Of course we could make a statistical "climate forecast"—a frequency distribution of the probability that the ball would hit any pin if we played the game a thousand times—what is sometimes called a Monte Carlo experiment.

So what has the analogy to say about projections of the impacts of human disturbances on climate? Ed replied, "What if we tilted the pinball machine?" We'd still have no more predictability than a few ball-pin encounters—weather unpredictability again. But the bell curve of the altered probability of ball-pin encounters would be perfectly predictable—that is, we'd have many more encounters where the table was tilted downward and many fewer where it was tilted upward. So the boundary value problem—the altered climate of the probability of various ball-pin encounters—can be predictable well beyond the initial value problem, as Ed explained it.

In terms of the second GARP objective, to improve the projecting of climate variability or change, we weren't completely clear whether that meant predicting months and seasons ahead or decades to centuries ahead—these projections occur from very different processes. For instance, if you had an El Niño, could you better forecast extra heavy rain in Peru or in California because of the El Niño conditions relative to a non–El Niño year? Probably so, we have found. And for the very long-term projections, couldn't we better forecast global temperature if we knew CO_2 was doubling from human activities over decades to centuries? With the ability of satellites to obtain data, and computers to process the data and run models over simulated periods of time as long as hundreds of years, it was becoming technically feasible to ask quantitative questions about century-long climate change. As a result, the GARP's second objective started to ascend in importance, creating an almost seamless transition from weather forecasting to climate forecasting among the world scientific community.

In the summer of 1974, GARP sponsored an international meeting to define the science that was needed for the second objective. John Kutzbach from the University of Wisconsin was put in charge of writing the report. John was a very broad-minded meteorologist, trained originally by Reid Bryson. Therefore, he had already learned how to ask broad questions, yet with his meteorological pedigree, he didn't frighten away mainstream meteorologists the way Bryson would have. Bryson's forays into anthropology and agronomy, and his unshakable belief that the world was cooling because of dust generated by the overgrazing and blowing soil from goats of Asia and Africa, were problematic for some scientists. Although the meeting wasn't multidisciplinary to the extent of including social scientists, it did include oceanographers and chemists looking at a broad range of issues, including stratospheric ozone depletion and air pollution chemistry.

The group also had a comparable representation of modelers and measurers. The modeling contingent was lead by Joe Smagorinsky, and the measuring group by French satellite meteorologist Pierre Morel and satellite pioneer Verner Suomi from the University of Wisconsin. Joe and Pierre were not shrinking violets, and they clashed constantly. Even the good offices of Vern, who tried in vain to maintain the peace, didn't stop what went on at the meeting. The climate modelers defined the precision requirements for the kinds of observations from satellites they needed, involving high-resolution, highly accurate global data sets. Immediately Morel said, in essence, "This much precision is not necessary. We can't build instruments to those requirements now. It's too expensive, and national governments will never support it. So if you claim that's what you must have, we'll get nothing, because they will say we can't meet that requirement." Morel's argument was at root an economic/political argument that modelers should not ask for precision beyond what the satellite technology was capable of delivering.

Smagorinsky then snapped, "We're not challenging the manhood of you measurers just because you can't measure up. This is what we need!"

I don't remember enough French curse words to quote the response from Morel, but it was very entertaining. Then Vern tried to jump in, saying, "Wait, wait—what we need to do is point out that what we can do, and what we will be able to do in the next five years, will improve the situation, even though we would like to go beyond that in the ideal world."

Of course, that was the right answer, and Kutzbach wrote that down with some relief. During dinner, I loved Vern's metaphorical description of the behavior we had just witnessed. "That was spreading land mines in an already treacherous landscape," he said, "and we need mine sweepers, not mine layers, if we are to get progress on these problems." Vern was one of a kind, and he had an apt metaphor for nearly every situation designed to get people unstuck from ego, narrow-mindedness, paradigmatic thinking, and tribalism.

The other major contention at the meeting was my fault. I had by this time been exposed to people in agriculture and ecology, so

I suggested that scientists interested in impacts of climate change needed to know how specific variables—such as drought and flood frequencies and temperature extremes—would change, because these have major impacts on agriculture, ecology, water supplies, coastlines, and so forth. Climate modelers were aghast—nobody believed we could produce credible estimates of such extremes. I was well aware that projecting trends in extreme events was not then very reliable, but I wanted to push the envelope to start refocusing research to make more-credible predictions sooner.

Basically, I had stirred up the same argument as that between Smagorinsky and Morel. The question isn't only what science we are comfortable producing given the limitations of the state of the art, but also what does the scientific—and yes, the political—world need from scientists, even if our methods are less than we'd like to make high confidence assessments? Later on, this tenet became known as "science for policy"—study what is needed, not just what is easiest to do. What was going on before that was really "policy for science"—how governments should fund satellites and computers we needed in science to pursue the second GARP objective.

I asked if I could have ten minutes to address the meeting and raise the question. I described the world food situation—food reserves by 1974 had dwindled to less then 10 percent of annual production, and food prices had skyrocketed. We were going to face famines if more severe weather variability continued, I said, and we needed to ask these questions. I stood alone in arguing that we had to consider the social implications of what we were researching. Soviet atmospheric turbulence expert A. S. Monin angrily attacked this idea as utterly irresponsible because we should not address any questions before we scientists were ready. He reminded me of Dick Lindzen. On the other hand, Bert Bolin was immensely supportive. He said that our science has to evolve, so that we address not only questions that we think are scientifically important now, but also questions that the people who fund us think are socially important to answer.

That was a prescient comment. As the general chairman of the IPCC 15 years later, Bolin would carry that open-minded philosophy with him. He waited a long time until the world was ready for somebody with his vision.

Paul Crutzen—the Dutch atmospheric chemist who won the Nobel Prize in chemistry in 1995 with F. Sherwood Rowland and Mario Molina for their work on stratospheric ozone depletion—attended that meeting, and like most of the others, he didn't say anything on the subject of science for policy in public. He came up to me later, though, and said, in essence, "You know you are fundamentally right, but it is going to be tough to do that quickly, so let's work our way toward it slowly." I was pleased that a few critical people agreed, and I understood why they didn't want to go on record saying it. In the early 1970s, arguing for the social implications of science that was still emerging was still viewed as suspect behavior by the bulk of the established community.

EARLY OZONE BATTLES

Paul Crutzen's work on ozone in the atmosphere opened up a whole new arena for climate controversies, which would continue unabated throughout the 1980s until the countries of the world finally agreed to stop ozone depletion In 1974, however, the U.S. Arms Control and Disarmament Agency asked the National Academy of Sciences to evaluate the environmental consequences of nuclear war. Mike Mac-Cracken and I were the two primary people assigned to the climate aspects. We examined scenarios where multimegaton bombs—a thousand times more powerful than the Hiroshima bomb—would be used. Long-range missiles weren't very accurate then, and you could not be sure of a direct hit, so you needed a massive explosion with a large fireball and broad footprint in order to take the target out.

These massive explosions would dump megatons of dust in the stratosphere. MacCracken and I calculated optical depths from the numbers that the Defense Department scientists gave us and concluded that the equivalent effect would be one hell of a non-natural volcano. The planet might cool a degree or so Celsius over a year or two, but cooling wouldn't be the main problem, merely an after-effect adding to the unimaginable misery.

Crutzen was working on the ozone question. He and his colleagues concluded that a tremendous reduction in ozone would occur, because 10-to-50 megaton nuclear weapons were so explosive that their fireballs were more than ten kilometers (six miles) in diameter. As a result, they would reach high into the sky, and they would dump the chemicals produced from the fireball in the stratosphere. Among the chemicals produced from that kind of heat are nitrogen oxides. Thanks to the earlier work of Harold Johnson of the University of California–Berkeley, and Crutzen's on supersonic transports, nitrogen oxides were known to deplete ozone. They calculated out a dramatic ozone reduction after a nuclear war, in the range of 50 to 80 percent, lasting for a few decades. That would cause lots of skin cancer, but I suppose skin cancer is a minor concern compared to billions killed by the direct blast and radiation effects, and hundreds of millions more dead as a result of the loss of infrastructure that provides food and health care and so forth.

In the next few years the field of climate science expanded into another thorny area, one that continues to have grave consequences to this day. I had continued to invite guests from all around the world to give talks at the Climate Club at NCAR. One day I got a phone call from John Holdren, then the energy analyst at the University of California, Berkeley (and now President Obama's science adviser and the director of the Office of Science and Technology Policy). John asked me to invite George Woodwell, the ecologist from the Marine Biological Laboratory in Woods Hole, who later founded the Woods

Hole Research Center. George had a controversial theory. He argued that the five billion tons per year of carbon that was injected by burning fossil fuels was being dwarfed by a factor of two or three by the amount of carbon produced from the burning of tropical forests. He was holding this position long before rain forest protection became a popular battle cry.

Wally Broecker, the original-thinking geochemist and oceanographer at Columbia University, considered this to be outrageous extrapolation. At a Department of Energy meeting in 1977, he screamed at Woodwell, "You have one data point in Venezuela, George—you've just extrapolated from Venezuela to the world!"

Since Wally had just published a paper in *Science* in which he used the ice core record at Camp Century, Greenland, and then extrapolated it to cover the world, I said, "Nobody here would extrapolate from Greenland to the world, would they, Wally?"

He smiled and said, "Okay, okay, I did it too, but he's wrong."

The key point is that nobody had enough data to really answer the question of carbon emitted from deforestation.

Norman Myers, an independent scientist from Oxford, was simultaneously calling the world's attention to the carbon emissions from deforestation. He was also concerned with the loss of biodiversity. His 1979 book, *The Sinking Ark,* was at the forefront of this movement. These environmental concerns were in parallel, and the communities began to learn about each other—the atmospheric scientists needed to understand what was happening in ecology, and the ecologists needed to understand the climate connection.

When George Woodwell gave his Climate Club talk, Paul Crutzen was in the audience along with a visiting scientist and colleague from the Max Planck Institutes in Germany, Wolfgang Seiler. Paul and Wolfgang were working on carbon monoxide (CO), and a prime source was deforestation-related fires. They were keenly interested in Woodwell's deforestation calculations, which would dramatically affect their understanding of the stocks and flows of CO, and they were stunned by his numbers. The figures didn't square with their CO observations, so they didn't trust them. But they could not completely rule out the possibility that the figures were correct.

So Crutzen did what any good scientist would. He organized an expedition to the Amazon to see for himself. The way he described their collection adventures, they were running in and out between burning trees, weighing the logs before a fire and after, bagging emitted gases, and so forth. The way the graduate students told the story, Crutzen sat in a Jeep directing them to run back and forth between the burning logs. Who knows what's true. . . it was good work.

In any case, they collected a massive data set on chemical emissions and processes, which led to two discoveries. A significant fraction of the carbon was not entering the air as either CO or CO_2, although most of it was CO_2. The carbon was being converted to charcoal, being dumped in the soil and into the air as soot aerosol. The flaming biomass produced a large amount of soot aerosol, accounting for some 25 percent of the carbon emissions. Paul was already beginning to reconcile his CO numbers, which indicated lower biomass burning emissions, with Woodwell's numbers, which were not based on chemistry but upon the actual deforestation rates in one place. Crutzen said, we can make up a quarter of the difference at least, maybe half, just by recognizing that burning those trees doesn't automatically convert to carbon dioxide in the air. The residue could be contained in soot aerosol and charcoal sequestered in the soils.

Crutzen would return to the nuclear holocaust question in 1981. By that time he was heading the Max Planck Institute for Chemistry in Mainz. The new U.S. President, Ronald Reagan, believed that the proper U.S. strategy to stop the Russians was not to meet them with conventional weapons, because they could outdo NATO in manpower and ground forces, but to threaten that if they came near Europe with ground forces, we would hit them with tactical nuclear weapons. Caspar Weinberger, Reagan's secretary of defense, gave speeches in Europe in the spring of 1981 in which he said that such a limited nuclear war would be won by the United States and its allies. This was essentially heresy, because until this point the doctrine had been MAD, mutual assured destruction, which meant deterrence, not fighting a nuclear war that would be lost by both parties. Weinberger was the first high-level person to say that we could actually win one of these wars, if it came to that. The media extrapolated from his statements that therefore it would be rational to consider having one. It rekindled almost overnight the antinuclear movement in Europe, and to some extent in the United States.

The Swedes, distrustful of the Reagan pronouncements and having always had a pacifist bent, immediately commissioned a study on the consequences of nuclear war, headed by Paul Crutzen, since he had already conducted the ozone depletion studies in 1974 for the National Academy of Sciences. Paul had a visitor from the University of Colorado at the time, chemist John Birks, and they both studied this question. Nuclear technology had radically "improved" from 1974 to 1981, and as a result, the nuclear arsenals of the superpowers, particularly the United States, were now filled with many more but much lower-megaton bombs. They could deliver greater, pinpoint-precision accuracy because of the new electronics. The size of their fireballs would be much smaller and not necessarily reach the stratosphere. Therefore, Crutzen discovered, these fireballs wouldn't be injecting that much nitrogen oxide into the stratosphere.

Paul remembered from his trip to the Amazon that a burning forest emits soot and many volatile chemicals. They calculated that from the emissions a nuclear war and all these fireballs would generate, and from the collateral fires produced as these weapons exploded at various sites around the world (and he was largely thinking of burning forests in this case), there would be so much junk put in the air that it could create a toxic photochemical smog. To create a toxic pall from the emissions of the biomass burning, sunlight would be required, because of its photochemical reactions that make smog. But the soot was going to block out sunlight, they lamented. Their new super-smog effect wouldn't occur, because the sunlight would not penetrate low enough into the atmosphere to create the photochemical reactions needed.

All of a sudden they had their eureka moment. That's the effect—nuclear explosions are going to generate aerosols that block out sunlight. And that's how nuclear winter was conceived. Science works in strange ways. You cannot always calculate what you are going to learn. Discoveries often are made by serendipity—but that only succeeds when brilliant people are assessing the results with minds open to unexpected possibilities.

CULTURAL ANTHROPOLOGY MEETS CLIMATOLOGY

The most famous anthropologist of the 20th century, Margaret Mead was known for her groundbreaking research and fieldwork in cultural anthropology. Her studies of childhood development and family structures in societies in Samoa and New Guinea revolutionized the Western conception of the social organization of indigenous communities. But her role in the arena of climate change has been little celebrated.

Margaret Mead was a powerful character and an important influence on my life's work. I got to know her through Will Kellogg. In 1975 Margaret was the president of the AAAS and very much the eclectic scientist, interested in a broad range of issues, interdisciplinary and innovative. Will brought me into a project that he had developed with her. Margaret had a grant from the National Institute of Environmental Health Sciences, in Research Triangle Park in North

Carolina, to run what was called the Fogarty Conference, sponsored by the Fogarty International Center. Most of those conferences discuss health issues, and Margaret decided that the subject that she wanted to explore was planetary health. Will ran the meeting with Margaret—he was the scientist, and Margaret was the visionary.

The conference had a catchy title: "The Atmosphere: Endangered and Endangering." Margaret wanted to assemble a diverse set of what she liked to call "analogical thinkers." She said she hated digital thinkers, people who only operated on the basis of methods. The people that she brought together ranged from G. D. Robinson, a conservative scientist who also had been at the SMIC meeting, to Wally Broecker, a brilliant and innovative maverick, and the enigmatic James Lovelock, who had a new idea called the Gaia hypothesis that he had developed with microbiologist Lynn Margulis.

Jim Lovelock believed that Earth was a negative feedback system that self-stabilized the environment for its own good. The Gaia hypothesis (named for the Greek goddess of the earth) integrated Earth's biosphere, atmosphere, oceans, and soil, and it had grown from Jim's earlier work for NASA on researching conditions for life on Mars—he found none. At its most basic level, the Gaia hypothesis asserted that the biosphere automatically operated global controls on the surface temperature, atmospheric conditions, and ocean salinity in order to stabilize all conditions in an optimal physical and chemical environment for life on this planet.

Lovelock was an independent scientist who was famous for inventing a sophisticated instrument that could measure chemical concentrations in the tiniest units, such as parts per billion or smaller. It enabled us to detect chemicals such as chlorofluorocarbons, and some of Jim's funding came from the chemical industry. The Fogarty Conference was one of the first opportunities we had to debate Jim's new ideas of Earth as an integrated organism, years before he published his 1979 best seller, *Gaia: A New Look at Life on Earth*. Taking the other side—arguing that humans were multiplying out of control and using technology and organization in a dangerous, unsustainable way—were the two Young Turks, 31-year-old plasma physicist John Holdren and me, the 30-year-old former plasma physicist, along with our more senior colleague, George Woodwell.

The charismatic scientist in charge, Margaret Mead, started the conference by saying, in effect, "If we can't solve the problem of the planet when it is as obvious as pollution, we can't solve any global problem."

As usual, certain people denied that we needed to solve a problem, since it wasn't proven yet to be very risky. The battle lines were drawn, and it was really a spirited meeting. When somebody said, "We don't have enough data to make any definitive statements," either Holdren or I countered with our usual battle whoop, "How can we take an unmitigated chance when we can't rule out catastrophic outcomes?"

Lovelock was among the most controversial speakers. His approach was, "Let's not worry so much about what we are going to do to the planet. After all, when the microbiota in the oceans produced oxygen two to four billion years ago, it was a poison that eventually—about two billion years ago—leaked into the air, relegating them to the anaerobic niches of the planet and making it possible for all the advanced forms to evolve on land. Therefore, life is very resilient."

Holdren and I jumped into battle mode saying, "Jim, we're not worrying about the survival of life on Earth. Rather, we are talking about the survival of half the species, and the well-being of human beings. We are not interested in the abstraction about whether DNA replication and evolution continues—of course it will. But there are four billion people on Earth tightly locked into national boundaries and a billion are at nutritional margins, so any environmental stresses on top of that precariousness could be catastrophic, even though not threatening the human species itself."

I'll jump ahead in time to continue the saga of the Gaia hypothesis, because it was an important episode in the climate science skirmishes. I thought Lovelock was brilliantly creative and I admired both his intellectual originality and beautiful writing style. About a decade after this Fogarty Conference, I was interviewed by the BBC *Horizon* science documentary team on Gaia. I was cautiously supportive of radical thinking, but said I found many contradictions in the theory that biotic interactions with climate were always good for the biosphere—for instance, the ice ages, where phytoplankton in the oceans increased their benefit by taking up CO_2 which amplified cooling of the planet, decimating the boreal forests. I preferred the concept of the "co-evolution of climate and life"—agreeing with Lovelock and Margulis that we should look at the planet as a coupled biophysical living system. I just disputed that negative feedback—maintaining conditions on Earth comfortable to life—was the only interpretation of paleoecological data, like the counterexample of plankton and forests just given. It sounded pretty Darwinian to me, not a radical new departure overthrowing old biological doctrine.

A few months later, I enlisted Penelope "Penny" Boston, NCAR microbial ecologist, to help organize the American Geophysical Union (AGU) Chapman Conference on the Gaia Hypothesis. We wanted to get this debate where it belonged—inside the world of mainstream science, not just on TV or in countercultural progressive magazines like the *Coevolution Quarterly*. It took more than two years to organize it, because many mainstream geophysicists on the AGU Board believed, as Wally Broecker said in a handwritten note: "I can't support a scientific meeting on Gaia since it isn't science! Cheers, Wally."

The opening was a tour de force from Jim and Lynn. But next up was Stanford's population biologist and close colleague, Paul Ehrlich. Paul, as I've noted, is a "take no prisoners" type. He began his talk—designed as a defense of natural selection—by attacking me for putting him in a position very uncomfortable for this iconoclast: "Darn it, Steve, you know I love heterodox positions but you have stuck me in a place where I have to defend orthodoxy." I knew something Ehrlichian was coming. Looking straight down at Jim and Lynn in the first row, Paul calmly declared, "There are about a hundred thousand examples of natural selection and essentially none for Gaia."

The meeting was a spectacular success—though scientific and philosophical challenges to Gaia were quite telling. Geophysicist James Kirchner and his University of California–Berkeley Ph.D. adviser, physicist-ecologist John Harte, wrote a brilliant philosophical dissection of five levels of published Gaia hypotheses all taken from Lovelock's and Margulis's writings. They said Gaia couldn't be all these definitions at once, implying the Gaians were very sloppy in their details and definitions.

Gaia was so loved or hated by different tribes that it was actually good for Earth systems science, in my view. Even though that passion sometimes obscured objectivity, people were spurred to work very hard at the analysis needed to understand how the Earth system worked.

Interestingly, Jim Lovelock now is a passionate believer that human greenhouse gas emissions can cause traumatic climate responses and is advocating eliminating coal burning and replacing it with nuclear power and other energy systems that don't emit carbon. Ironically, when I spent a weekend with Jim and his wife, Sandy, at Big Sur recently, I was in the strange position of trying to moderate his more apocalyptic recent views—what a change of roles from our first meeting!

In 1977, several years after the Fogarty Conference, I invited Margaret Mead to be the commentator on a panel about how to work effectively in an interdisciplinary model when it is embedded in a disciplinary shop—such as a university or a national lab. She had generously agreed to be a member of the editorial board for the radically new interdisciplinary, peer-reviewed journal that I was launching, called *Climatic Change*. It has had a successful run for more than 35 years and is still making a strong contribution, I'm happy to say.

During the panel presentations at the 1977 AAAS meeting in Denver, Margaret took the microphone to give a summary comment.

"How do you do anything creative in a hidebound elitist institution like a university or a national lab?" she asked, and then she told us. "It takes three things to create an interdisciplinary group or a new idea. First, a charismatic leader—and that charismatic leader has to have enough credentials in the disciplinary world to fight off the narrow minds. Number two, attract a staff of young, brilliant analogical thinkers. The digital thinkers will kill you. The methodologists will drive you nuts. You want people who at the margins of their intellect want to solve problems and want to make connections. And the third thing you need is a sugar daddy. You need somebody to come up with the bucks, because you are not going to get them out of the establishment."

Her assessment was very astute. Most of the creative institutions I've known falter when their charismatic leader departs. What is required to institutionalize creativity to take it beyond the charismatic leader? If we look back to the early days of NCAR when we began to develop climate research, I think Margaret Mead's model works. There were charismatic leaders—Phil Thompson, Bob Dickinson, Warren Washington, Will Kellogg, Ralph Cicerone, Paul Crutzen, and even me. We were able to broaden the vision to the point that it became institutionalized. The sugar daddy was the National Science Foundation, and we had a mix of analogical and digital thinkers. To make climate studies work, you can't just have people who think at the margins about connections—you also have to have people who can build the models, run the data, and so forth. So maybe Margaret was a little too harsh on her second point, at least after the start-up phase.

Margaret Mead was one thoughtful, feisty old lady. Just before the Fogarty Conference in North Carolina, I had gone to visit her at the

American Museum of Natural History in New York. She worked in one of the Victorian towers at the very top on the fifth floor. I had to be escorted through the back rooms of the museum, past all these old cabinets stuffed with dead birds and other carcasses. There may well be species buried there that nobody even knows about, waiting to be discovered in the collections.

Margaret was at first kindly, as we talked about her upcoming meeting on the planet's health. Then I told her about my *New York Times* story with the Mark Twain quip, how someone had stamped "bullshit" on the copy displayed at NCAR, how my promotions were being threatened, and how a prize I had been nominated for on my cloud feedback paper was even withdrawn because of my public persona. I guess I was expecting this grandmotherly lady to put her arm around me and tell me that I was an unfairly treated, good boy.

Not a chance. She said, "Listen, kid, if you can't stand the heat, get out of the kitchen." Then she gave me some advice that has stuck with me for more than three decades. "That's exactly the way it's going to be. You are going to have to deal with this, because these people are being threatened—because you're saying the way they define quality and the way they see their lives is not the only way, and that's not going to float for many of them."

And then she smiled. "Have a generational perspective. Don't just think about it from day to day; it will change slowly over time. Your students and theirs will not be locked into their paradigms. They will have minds to make up for themselves. Things will change. Be tough! Hang in there, stay true to principles, hang around with like types, and you might even live to see an improvement in the world that you had something to do with—even if it takes a whole career."

I'm still hanging in, slogging in the trenches of the climate wars, and so are many of my early colleagues, including John Holdren, Jim Hansen, Ralph Cicerone, Paul Crutzen, V. "Ram" Ramanathan, Mike

Schlesinger, and Richard Somerville. The sparks we generated in the mid to late 1970s were about to ignite at the First World Climate Conference in 1979 and at the first U.S. Senate hearings on climate change as an energy policy debate. We were moving beyond the halls of science onto the world stage.

DEEP INSIDE THE BATTLE TO SAVE EARTH'S CLIMATE

3

EVEN AS THE SPECTER of global warming loomed closer, the ability of climate science to uncover its peril grew steadily more sophisticated. The nascent awareness among scientists in the 1970s was replaced by an outreach that spread from television shows to the first world conference on climate change. Along with growing public awareness, though, came confusion. Beyond conservatives in the realm of science, the Reagan Administration stoutly opposed any efforts to rein in polluters. That was to them anathema to the free enterprise system. Nor was our message always unified, a problem placed on full display in a dispute between Carl Sagan and me over the prospect of nuclear winter. So while our voices were increasingly being heard, the ground was also being laid for the naysayers to use our own words against us.

By this time I had no doubts as to where the indicators were pointing. In *The Genesis Strategy: Climate and Global Survival,* my first book, I predicted that greenhouse gases and perhaps aerosols would cause a "demonstrable" change in climate by the end of the century. Published in early 1976, *The Genesis Strategy* explains what we had learned about the global implications of measurable CO_2 and other greenhouse gases and aerosols, based on constantly improving observations and models, and what might occur if the world continued in that direction. The "Strategy" part included suggestions and recommendations to avoid potential catastrophe. The "Genesis" part alluded to the story in the biblical Book of Genesis where Joseph interprets Pharaoh's dream that seven good years are to be followed by seven lean years; the strategy to combat adversity is to take proactive precautions. I argued for adequate food reserves as a hedge against unfavorable weather variability, but I extended it to include preparations to deal with a likely future of climate change. We had to be ready for that too—mostly by developing alternative energy supply and demand technologies that emitted less greenhouse gas.

The book had a substantial influence on a number of young scientists, and several of them applied to the Visitor Program of the Advanced Study Program at NCAR—which I directed for nine years, starting in 1981—to pursue climate science. I didn't find out until after they arrived that it was because of the book—and even more surprising, that they had discovered the book by watching late-night television.

Paul Ehrlich and Carl Sagan both had said independently to Johnny Carson, host of the *Tonight Show,* "You shouldn't just have us as your two scientists on the show, you should get this guy Schneider." So in the summer of 1977 I appeared on four Carson shows. Steve Allen was the guest host on my first one, and it was the thrill of a lifetime to sit there with this clever, funny guy I had long admired to discuss in popularized terms why the public should care about atmospheric science. The next three programs with Johnny really turned out well. Viewers didn't know that Carson's guests worked out the questions in advance with the producers, and they gave the producer the answers too. The first time I went on air with Johnny, he would ask a question, I might get halfway through the answer, and then he would sort of hesitatingly stop and say, "But wouldn't it be . . . ?" And he got the answer, and I would say, "Yes!" The audience would be amazed at his perspicacity and intuition.

He was a smart guy, and he actually did catch on quickly, but the ruse was the ultimate showbiz trick. Even though we didn't stage it or modify the truth, it was clearly entertainment, not a news program. Johnny and I got on very well. They were good programs—the network and the audience liked it.

Yet my spot on the *Tonight Show* abruptly terminated in September 1977 because I made the ultimate mistake, unwittingly. Johnny and I were chatting on air as usual, and I had prepared slightly different questions that night. Johnny started going back to earlier questions we had done before, and I didn't like the idea of repeating stuff, because I had already been getting some comments from my atmospheric science colleagues: "Hey, you are saying the same old stuff. Can't you say something new from program to program?" So I changed the format, and off the cuff I addressed the audience. "Let's take a vote, ladies and gentlemen. Let's see if you know—how many of you think the world is cooling?" Most people thought the world was. "And how many people think it's warming?" Few hands were raised. I recall saying, "No, the world actually is warming, it's just that we had a short-term cooling trend."

I continued the Q&A with the audience, Johnny participated, and I thought everything was fine. At the end of the program, though, when we shook hands, he walked away abruptly. The producer came over and said, "Steve, that was a brilliant show, but it will probably be your last."

"Why? Wasn't it great?"

"You deviated from the script! You took over his program—that's not for guests to do!" The producer shook his head at my stupidity.

I was never called back. However, when postdoctoral fellows told me that they became interested in climate because they saw me on Johnny Carson and then read *The Genesis Strategy*, I decided that the showbiz summer was worth every minute.

This media outreach did not occur in isolation. As we expanded the interdisciplinary study of climate change, some of us recognized how critical it was to inform the general public of our findings. I welcomed any opportunity to put the research before the people, after proper scientific diligence on how much confidence we had in it, but I was in a minority. Scientists in general are reluctant to pursue publicity in the wider world, preferring to make their reputations in peer-reviewed journals and other publications within the scientific establishment, where information is both conveyed and commented upon by people with expert knowledge. There are no such controls in the public media. It's a debate that continues and has even accelerated in this world of bloggers and Twitterers, where misinterpretations of data can travel around the globe instantaneously.

Because of the media and the book, I got a phone call from CBS. Dan Rather, a contributing correspondent to *60 Minutes,* had an idea for a new show. He was going to do a *60 Minutes*-like program, but it was going to be biographical. It would be called *Who's Who.* They had taped a couple of shows already, and the fourth program was to include Lily Tomlin, Rosalynn Carter—and me. What better platform than to be sandwiched between one of the First Ladies of comedy and the First Lady herself?

Dan Rather and his producers arrived in Colorado at the height of the 1977 drought. I had written in *The Genesis Strategy* that we'd had a long lucky streak with the weather, and we had to expect nasty weather one day soon. I did not predict that we would have a drought that year. I simply said we were due, because we hadn't had one for a while—and guess what, it happened to occur in that year. So I was given credit as a prognosticator that I did not deserve, and I didn't claim the credit, but I got it anyway.

We were filming a conversation at NCAR when Rather said, "Haven't we got anything more visual than you talking here with the mountains in the background?"

I told him about a lake up in the mountains with a water level so low that viewers could see how far the drought had drained it down. "Why don't we go fishing?" I said.

I gave him one of my two fishing rods and we headed out. The camera crew set up about a hundred yards away, and we were wired with microphones. It was a sunny day, so I suggested using a copper Kastmaster lure.

Dan said, in that famous anchorman voice, "I've been fishing, son, since I was young. I know what kind of lure to use."

He chose a silver lure and I put the copper one on. As we fished, we talked while the camera was filming. About ten minutes later, I caught a fish. Dan said, "Oh, you've got one." He seemed miffed because he wanted to catch the fish—it was all in good fun, of course. I reeled the fish in, and it was squirming around on the end of the hook. Then we cut the scene. Dan yells out to the guys, "Did you get that?"

"No, we were changing the film."

As instructed, I cast the fish back in the water still breathing. When the camera was reloaded, I reeled in this poor fish, which had no fight left in it. This virtually dead fish coming up on the line was what they put on the air. The irony is that I was never much of a fisherman. The true fanatic, John Holdren, had taught me how to fish in the Colorado lakes only the summer before, and to use the copper lure in that lighting. When John watched this program a month later, he called me. "What did you do—put a dead fish on the end of the line?"

THE EVOLUTION OF INTEGRATED ASSESSMENT

I was hardly a media darling, however. Most of the time I was still pursuing my scientific studies. In 1975 I became founding editor of the new journal *Climatic Change*—which I initially founded, truth be told, to spite my institute director, who had warned me that if I didn't stop this "interdisciplinary bent," I might not get promoted to tenure. It was an honest warning of the prejudices of the promotion committee, but I was so annoyed that I set up *Climatic Change* as an interdisciplinary peer-reviewed journal to show the world it could be done with scientific quality, regardless of their preconceptions. I was working to broaden the climate science paradigm to include multiple disciplines and to provide a forum for research and discussion of interdisciplinary solutions.

The larger world of atmospheric science was beginning to recognize that part of its next phase would be as a partner, particularly with experts in economics, ecology, agriculture, oceanography, and hydrology. In a similar spirit Bob White, who had been the head of NOAA, set up the Board on Atmospheric Sciences and Climate (BASC) through the National Academy of Sciences. He organized summer workshops, two to three weeks long, at the academy's Woods Hole retreat center, in which he tried to define the new evolving direction of climate research.

Bob invited me to the first workshop in 1978. I was nearly the only one—surrounded by mostly traditional meteorologists and atmospheric scientists—arguing that we needed to have a multi- and eventually interdisciplinary approach. What passion that seemingly self-evident idea generated from the assembled scientists. These were not contrarian ideologists, but keepers of the disciplinary faith.

The following summer, Bob White added Mickey Glantz, and renowned geographer Ken Hare from the University of Toronto, and Norman Rosenberg, an agro-meteorologist from the University of Nebraska, to the group. That time around, four or five of us fought to include integrative, interdisciplinary components, and the effort wasn't so controversial anymore. People have to get used to change, I learned, and Bob was brilliant in his capacity to develop group consensus slowly over time.

One fascinating addition to the interdisciplinary approach extended the scope of study far into the past. I had spent a few months in 1976 at the Lamont-Doherty Geological Observatory with Wally Broecker and Jim Hays, a senior paleoclimatologist who had just coauthored a famous paper on the frequency statistics of ice ages. In 1982, as a UCAR affiliate professor, I became immersed in geology, learning all about paleoclimate and the need to use the paleoclimatic record as the backdrop against which to calibrate our understanding of important processes. It also could help validate the tools that we used to make predictions of the future.

We don't use paleoclimate—that is, prehistoric situations—to give analogies to the future. That doesn't work well because land use changes, atmospheric CO_2, and aerosol forcings will be the main drivers of climatic changes in the near future. Those factors are all going to be different from anything that's ever occurred in history. Historic information can be employed to study processes that help us to build the models we use for the near future and to test them. The charismatic leader of that tribe, John Imbrie at Brown University, put together a vast team to reconstruct the temperature patterns of the last ice age some 20,000 years ago, and they produced maps—not perfectly coincident with today's understanding, but remarkably close for a first attempt.

The process began to emerge, as the development of integrative studies and climate was going to be defined much more broadly than as atmospheric science alone. We had the venue of *Climatic Change* to publish these interdisciplinary papers.

With Bob White and the Board on Atmospheric Sciences and Climate established at the NAS, we now had official support. NCAR director Francis Bretherton began to work with NASA in developing Earth systems science. We brought Eric Baron to NCAR, a young student from the University of Miami who was very interested in climate and had been doing paleogeographic reconstructions of where the continents were in the Cretaceous period, when dinosaurs patrolled some hundred million years ago. Eric wanted to figure out what that meant climatically. So Eric, Starley Thompson, and I began to incorporate geologic information into our multidisciplinary perspective. I think it was really NCAR at its finest hour—helping to create a new direction for the Earth science community as a whole, while at the same time improving our atmosphere-related research products.

Our Climate Sensitivity Group was expanding our thinking in various ways. We were showing that CO_2 was likely to be doubling, and we were predicting the outcome a few centuries in the future, when the system comes into equilibrium—that is, settles down and is no longer changing. Yet I realized that these equilibrium results didn't matter as much as studies that showed the process developing over time. Ecologists and agriculturalists didn't care much about whether the temperature warmed up two degrees Celsius in 2300—they cared about whether it was two degrees in 20 years, in 50 years, or in 100 years. The rate of change was at least as important as the total eventual amount of change, since the capacity of a system to adapt depends on both the amount and rate of warming.

What really mattered was the transient—the time frame of the changes and the rate at which they would occur. Starley Thompson and I began to think about this problem. And with Turkish physicist Nuzhet Dalfes from Rice University—doing his Ph.D. with me at NCAR—and Starley and Eric, I also set about making homebrew wine. We would have an ethnic food party about every other week. Bob Chen from MIT was working with us as well, and he would make dim sum. Nuzhet would make baklava and Turkish food, and Starley would have to do some kind of Texas ribs. I got this whole winemaking apparatus, and we were scrunching grapes and making plum wine from backyard plums. And, of course, while we were doing this, we were talking science, especially the right directions of our evolving field.

Later, drinking some of the very bad wine that we made, Starley and I noticed that this transient issue was more interesting than we had realized. We learned from ecologists and agriculturalists that the rate of global average temperature change isn't all that matters, because the middle of continents are going to warm up much more rapidly than the middle of the tropical oceans will. The mixed upper layer of Earth's oceans, typically a hundred meters thick, would respond within the time frame of years, whereas the middle of continents would respond in weeks to months. Then we considered the high-latitude oceans, where the cold and salty water sinks to the bottom, and the upper layer is hundreds of meters deep or greater. Its warming time could be take decades to a century. That means the middle of continents warms up rapidly; the middle of oceans an order of magnitude slower; and the higher-latitude oceans an order of magnitude slower than that.

During the transition toward a new equilibrium, the temperature difference from land to sea and Equator to Pole would be skewed. If you skew the temperature difference, you skew where the storms are. You skew the nature of the circulation systems. We turned to each other and said, "That's the problem."

The CO_2 problem and its potential impacts could not be confidently handled by equilibrium calculations any more—everything else must be a transient. This step toward integrated assessment evolved with students during long discussions in the sort of atmosphere where I encouraged everyone to guess what each other's computer model was going to say the next day when we got our computer runs back. There was no such thing as a "wrong" answer. I used to say, "The only bad question is the one you thought about but didn't ask!" We all laughed together when we guessed wrong, and we all strutted a little when we were proven right. Meanwhile, we were training our intuitions on how our tools worked—or didn't—and why. So the science continued to make discoveries even as we tried to alert others to the peril we had long ago seen in broad outline. Because science assessment is truly a collective effort to weigh and integrate individual contributions, the entire international community was becoming aware that climate change was far more than just a sideline in meteorological research.

WHOLE EARTH CONCERNS

In February 1979, the First World Climate Conference took place in Geneva. Sponsored by the World Meteorological Organization (WMO) and a number of other international bodies, it was one of the first major international meetings on climate change. Bob White was the co-chair with E. K. Fedorov of the Soviet Union. Even in the planning stages, we knew it was going to be a controversial meeting.

The beginning of confrontation rather than friendly discussion over climate was starting. The First World Climate Conference drove home just how much passions are inflamed by this subject. First, Bob White raised the question of whether this should be a meeting of government leaders to discuss policy to deal with climate variability and change. The rest of us thought that was a good idea, if the meetings were carefully structured and the agenda was relatively benign and agreed to in advance. Sir B. J. (John) Mason, head of the Meteorological Office in the U.K., and a number of others insisted that a ministers' meeting was at best premature and at worst irresponsible. How dare we think about policy for a problem that had not yet been scientifically solved! It was a mantra I'd heard before—it brought to mind the debates at the Climatic Impact Assessment Program (CIAP) with Dick Lindzen and Mike MacCracken seven years earlier.

When the WMO first decided to do the conference in 1977, the purposes were clearly stated. The attendees would "review knowledge of climatic change and variability, due both to natural and anthropogenic causes; [and] assess possible future changes and variability and implications for human activities." An issue of special importance was "the problem of possible human influence on climate."[1]

When the actual conference took place in February 1979, Bob White in his keynote address tackled the issue straight on, saying, "In recent years we have come to appreciate that the activities of humanity can and do affect climate. We now change the radiative processes of the atmosphere.... The potential consequences of increasing atmospheric CO_2 resulting from fossil fuel combustion are already a major concern.... The implications of further projected increases [in CO_2] are uncertain, but the weight of scientific evidence predicts a significant global surface temperature increase."[2] That statement could still be made today—and Bob said it three decades ago.

Scientists from a wide range of disciplines showed up for this historic meeting. We had plenary sessions and four working groups to examine climate data, the identification of climate topics, integrated impact studies, and research on climate variability and change. Of the overview papers, I particularly remember the one given by Sir John Mason.

He had just published a paper in which he had said that the atmosphere is resilient and is "wont to make fools" of those who don't understand the resilience, implying that human activities would not cause much climate change. He spouted off this palaver while strutting back and forth across the stage, and at the end of the show I raised my hand. I was really angry, because I had privately talked with him earlier about the numerous climate sensitivity experiments that we had already performed. Faced with that data, he could not possibly argue that there was no sensitivity of the climate to the heat trapping associated with greenhouse gas buildups to that point. We already knew that this amounted to an addition of heat of several watts per square meter, what climate scientists called radiative forcing. V. "Ram" Ramanathan, then at NCAR, had shown clearly how that forcing was distributed between surface evaporation, downward re-radiation, and so forth. He had also shown how the warming evaporated more water and further increased the moisture content of the atmosphere. We understood this basic physical picture pretty well. Although considerable quantitative uncertainty existed over how much CO_2 doubling would warm the climate, a few degrees seemed to be the best guess—not "nothing much," as Mason asserted with no supporting analysis.

I raised my hand. "Professor Mason, I'm sorry I didn't know the atmosphere is so resilient to forcing, because if I had, even though this is February, I should have brought my bathing suit instead of my skis."

"Well, Steve, you would be a bad climatologist."

"No, maybe the other point, John, is it's not resilient to forcing, because there's a hundred watts of energy per square meter difference between winter and summer, and the fact that winter is cold and summer is warm is exactly the proof that the system isn't resilient—that the system responds to radiative inputs. What we have left to do is figure out quantitatively what's going on."

Mason also said that we used to think SSTs were going to be an ozone depleter, but now we knew from new data that they slightly increase ozone. He joked, "Wouldn't that be an expensive way to maintain the ozone layer—*ha ha ha!*" That wasn't true either. The proposed U.S. Boeing SST was to fly higher than the Concorde, and preliminary calculations suggested not ozone increases, but decreases.

I stood up again, "But, John, you correctly mentioned that the complexities of nonlinear dynamics and chemistry made the calculations of Concorde SSTs go from an ozone reduction to a slight increase, but you forgot to mention that those same complexities made the original calculations of small fluorocarbon-induced ozone decreases into very large ones. Why don't you be symmetrical and say that when we fool around with the system without understanding, some things are going to get better, and other things are going to get worse?" Those battle lines later on became the contrarian debate.

On the whole, this meeting was well structured, despite the disagreements smoldering under the surface. So I was taken by surprise when

Fedorov from the U.S.S.R. in his opening speech delivered an ideological blast, claiming that there was a world food crisis, which was partly due to climate fluctuations that have induced these changes, but the underlying cause was not climate—it was the "manipulation of monopolistic capital" by—guess which country. I was sitting next to Mickey Glantz, and I remember muttering something to the tune of "That son of a bitch!" The most monopolistic capital bloc ever manipulated in agriculture trade was the Soviet purchases of grain in 1972 and 1974, driving up the prices and starving people in India who couldn't afford to compete on the open market. I started to raise my hand to nail this guy, but Glantz grabbed it. He said, "No, wait, wait, this is the co-chair—you have to give him a chance to bluster, he is a Soviet." His point was well taken, because at a later reception, while I was still fulminating about Fedorov, Bob White said, "Fedorov has to say that. It was the price of his going to the conference. If he says it again, skewer him, but he probably won't." He never did; Bob and Mickey called it on the money. We didn't have to have the confrontation, and I learned a good lesson about keeping your powder dry, especially at international gatherings.

When Ralph D'Arge—the pioneering economist from the CIAP days—gave his talk on economics, Mason and others excoriated him in the nastiest tones for daring to calculate the costs of climate change when "we haven't got a clue what the climate change is going to be." Mason missed the real clue: that you need to begin the multidisciplinary process of asking the question—"So what if the climate changes?"—in order to set up the question about whether or not we want to take the risks, long before we have the final definitive answer. And what's the probability of "definitive" anyway? It's a stupid word that has no meaning in the context of climate change, because few measurements or conclusions in complex systems are ever definitive.

The First World Climate Conference concluded with a formal declaration, which summarized how climate change might impact human activities such as forestry, fishing, agriculture, water use, and urban planning. The declaration made an appeal to nations "to foresee and to prevent potential man-made changes in climate that might be adverse to the well-being of humanity." But in the end, the conference ducked a recommendation that a conference of government leaders should be convened to take necessary international actions. A decade more of study—and more bickering—was to precede any action-oriented meetings.

FIRST U.S. HEARINGS ON CLIMATE AND ENERGY POLICY

We all know that the United States holds the number one slot in total global CO_2 emissions accumulated over time, even though a few years ago China became number one in the dubious contest of most carbon emissions per year. We also know what we ought to do about our emissions. Sad to say, the United States had the same options to choose from in 1979, when Connecticut Senator Abraham Ribicoff held the Senate's first hearing with climate change as a prime factor in the energy policy debate. He was joined by Senator Edmund Muskie of Maine, whom Ribicoff considered the number one environmentalist in Congress. Muskie had been a former member of the committee and had enlisted Roger Revelle to comment on future possibilities some years earlier, when Revelle was defining global warming and working on world hunger, even before the wake-up call of the energy/food crisis in 1974.

The nation was experiencing another energy crisis. People were tired of waiting in long gasoline lines. In some states they were permitted to buy fuel only on odd or even days of the month, rationed by license plate numbers. The accident at the Three Mile Island nuclear reactor in March 1979 had made people mighty nervous about expanding that form of energy as a fossil fuel replacement, and the senators and representatives heard plenty of complaints when they went home for the long Fourth of July recess that year. Senator Ribicoff had announced in

May 1979 that he was retiring after a long and successful career, most recently as chair of the Senate Committee on Governmental Affairs. Since he would not be running for reelection in 1980, he had nothing to lose by raising the controversial subject of energy policy.

I was invited to speak at the hearing on July 30, 1979; the symposium was entitled "Carbon Dioxide Accumulation in the Atmosphere, Synthetic Fuels and Energy Policy." Joe Smagorinsky preceded me with a detailed description of the state of the art in climate experimentation and its uncertainties, both quantitatively and qualitatively. He also suggested that given uncertainties in the science it was premature to consider policy—a traditional view that is still held by some scientists even today. I wanted to bring in a more general set of climate issues, concerning not only the question of low CO_2-emitting energy and development, but also the question of how societies deal with natural climatic variations.

I focused my testimony on what might happen if greenhouse gas accumulation changed the climate in ways that produced droughts and changes in rainfall, especially in countries where the population only marginally sustained itself as it was. The huge effect on food supply had recently been demonstrated by the famine in India caused by drought and the grain shortage, combined with Soviet wheat purchases at a massive scale that set off worldwide food price spirals. I also addressed water supplies and energy demands, in the face of a growing population.

Revelle, in an aside related specifically to agriculture, suggested that releasing carbon dioxide at levels that warmed the planet might be useful to agriculture in mid-latitude granaries. Growing seasons could be extended, for instance. But he also focused on the concerns of large increases in CO_2 on other sectors. Synthetic fuels could help reduce our dependence on foreign energy sources and perhaps one day prevent more gas lines—which is why President Carter proposed it in the first place. However, per gallon of fuel created, synfuels such as oil shale produce much more CO_2 than burning fossil fuels emit.

The topic of synthetic fuels was also a hot potato at this time. What would these fuels do to the atmosphere? And what about international issues like deforestation? All the key issues were brought up 30 years ago.

Roger Revelle suggested developing alternative sources of energy:

"We need a sizable commitment to a major synthetic fuels program at the present time, but also we should devote considerable effort to developing alternate sources of energy, which will not enhance the carbon dioxide effect on the atmosphere—I am thinking about the various forms of solar energy all the way from development of fast-growing trees for production of biomass fuels, photovoltaics, solar towers, solar heating, and I would argue also for nuclear, . . . at least development of the safety measures that are essential if we are going to make nuclear power acceptable. Also, hydropower and perhaps something like ocean temperature energy conversion. . . .. I think that all of these things should be part of a balanced program."[3]

Ribicoff retired, replaced by a young Chris Dodd as Connecticut's junior senator, and Ed Muskie went on to become secretary of state for the remainder of Jimmy Carter's term. Not much changed in policy terms, despite a recommendation by Ribicoff that the problem of CO_2 had to be considered as a global issue and solved with global cooperation. The Department of Energy (DOE), under Carter, developed from the Energy Research and Development Administration and was assigned responsibility for an analysis of the current trends in global climate change. Unfortunately, that effort would be vitiated by the Reagan Administration.

Two years after the Ribicoff symposium, as the deputy director of the Climate Project at NCAR, I testified at a new set of hearings organized by Albert Gore, Jr., the young Democratic congressman from Tennessee. A decade earlier, Revelle had been a mentor to Gore, when

Revelle was teaching about environmental issues at Harvard. Gore wanted to hear from expert witnesses in the debate over how the government ought to perceive the problem of carbon dioxide emission and what research it should support to assess potential damages.

The House of Representatives conference room in the Capitol was sparsely filled with reporters in July 1981—unlike hearings that Gore would conduct when he moved to the Senate. Few folks really cared that the new Reagan Administration was decimating the budget for studies of how climate change could harm nature or the economy, choosing instead to focus its impacts research only on potential benefits of CO_2 emissions, such as fertilizing crops. Interdisciplinary research was just starting in the United States at large scale with the DOE project headed by Revelle to study climate impacts, and the proposed Reagan cuts would set back progress by many years. Al Gore recognized this and helped set up the hearing as part of the congressional oversight responsibility.

I started out with an overview of the uncertainties of the impacts of the global climatic response to CO_2 increase and possible scenarios, ranging from minimal impact to the possibility of the catastrophic. Our research work was aimed at the all-important question: "So what if the climate changes?" Gore asked me to focus on what would happen to the United States, for the benefit of the policymakers on the committee. So I brought up the question, "Is it important who wins and who loses?"[4]

Supposing one part of the nation would benefit greatly and another part would be hurt, but there would be no net national change, does that mean there would be no problem? I doubt it, because there is a political question coming from that equity stress. Trying to come up with definitive answers from what is now known is absurd, because all you can do is try to build a scenario and find out where the critical points are.

Some people claimed that climate change is an intractable problem—that there isn't very much we can do about it. As I mentioned earlier, I believe there is a hierarchy of responses that you can consider. "The first one, and it is a policy," I told them, "is to do nothing, and that is not one that is widely advocated—except by some special interests." I continued:

The second one, which is a little more radical but not very, is the one we have been doing, and that is to study the problem. After all, studying is a policy decision, we have decided to try to invest in research to try to get more information so any policy we take in the future can be made on a firmer basis. . . . If we go back and say we even have 20 or 30 years before the CO_2 effect is clearly detected, we are still going to face a risk. That is, we will have to adapt to a larger dose of carbon dioxide and its consequences, good or bad, if we wait until we are sure, than if we act now and reduce the carbon dioxide burden.

The next most active policy step, which is the one I personally advocate, is to try to what I call 'build resilience' . . . namely, do whatever you can now to increase the number of options you will have available in the future so as the information comes in and the problem becomes clearly a problem or not, you have the flexibility to move.

That 1981 Gore congressional hearing continued over the course of many hours. When the last panel was convened, the proceedings took an ugly turn—the first time I had seen such ugliness in a climate hearing since I began testifying some five years earlier. Witnesses from the executive branch, who were appointed by the new Reagan Administration, stonewalled Gore's efforts to inquire into credibly rumored cuts in the Department of Energy (DOE) budget for its carbon dioxide program associated with climate impacts research—the Revelle effort that had been under way for several years. In his testimony, N. Douglas Pewitt, acting director of the DOE's Office of Energy Research, denied the budget cuts, although he admitted some "changes" were being made in the management of the program.

Gore pressed Pewitt for specifics. "Let me just ask you, in fiscal year 1982 in the area of effects of climate change and carbon dioxide increase on the environment, you still plan to spend $3.324 million?"

Pewitt replied that he intended to spend an appropriate amount of money, still to be determined. Even though Gore pointed out to him that the figure appeared in the National Climate Program Office document prepared in January 1981 by the DOE's assistant secretary for environment—and was not "hallway gossip," as Pewitt called it—Pewitt maintained that the budget for the effects of climate change and carbon dioxide increase was under review "to see if it is sensible."

A tiny fraction of the proposed Revelle/AAAS effort for the DOE involved a serious study of what kinds of information attracted the attention of Congress. Pewitt held it up as an example of spurious research funding. In fact, it was a sensible approach, for if scientists could not deliver their findings in a way that could be understood by members of Congress and perhaps influence their behavior, millions of dollars in research would be wasted.

Pewitt was using a tactic similar to what Democratic Senator William Proxmire from Wisconsin used to do. From 1975 to 1988 Proxmire held a press conference every now and then to denounce some small bit of research funded by the National Science Foundation or other agencies, giving it his Golden Fleece Award—ostensibly for fleecing the public, in his view. He determined this simply by reading titles of obscure-sounding research projects (and yes, some jargon-prone scientists bring this upon themselves). The good senator did not, of course, perform the due diligence of reading the proposals carefully to see why the research was proposed and why it was funded after passing rigorous peer review. Proxmire meant to convey the trappings of responsible senatorial oversight—while in fact he was merely pandering to shallow prejudicial thinking in an anti-intellectual framework.

Pewitt was carrying out the directive from his boss, Energy Secretary Dr. James B. Edwards, a dentist from South Carolina appointed by Reagan primarily to dismantle the Jimmy Carter–initiated, technically complex Department of Energy. (Remember Carter's famous aphorism "Energy is the moral equivalent of war.") One of Reagan's first acts after inhabiting the White House was to remove, at taxpayer expense, the already-paid-for solar collectors installed by Carter—to serve as a symbol of how little the newly elected president thought of renewable energy.

That contempt was apparent at an internal governmental briefing a bit earlier. As David Slade—the far-sighted manager of the Revelle program for DOE—proudly summarized this innovative and cost-effective program for the new DOE officials and other government people, Edwards leaned over to Pewitt and said, loudly enough to be overheard in the row behind (an "ear witness" told me), "We were put in here to get rid of environmental and social science research, so just forget this project."

Gore had called the hearing in the first place to head off the ideological meat ax of the Reaganites at the pass. He failed, of course. In the budget Reagan proposed for fiscal year 1982, he asked Congress to approve an increase of $3.8 billion for research and development for the Department of Defense, while deepening the cuts to the DOE, which supported the national research laboratories, by more than $1 billion. Reagan failed to dismantle the DOE altogether, perhaps because it had taken over the problem of disposing of nuclear materials from power plants and weapons.

In calling together the experts, Gore had expressed a sense of urgency in determining whether or not this global warming problem was in fact occurring. He asked Pewitt whether he shared that urgency.

"No," Pewitt said. He brought in the "alarmist" argument. He talked about the earlier examples of the aerosols and the SSTs disturbing the ozone layer in the atmosphere, which turned out to be less serious than anticipated, and said that more scientific research probably would not justify the more alarming CO_2 predictions. "I absolutely refuse as an official in a responsible position to engage in the type of alarmism for the American public that I have seen in these areas time and time again." The ozone "alarmism" Pewitt mentioned turned out to be much worse than scientists had forecast when the ozone hole was discovered in the mid-1980s.

Gore took a few steps over to a chart documenting carbon dioxide levels that had been gathered over the past 23 years, remarking that the chart reflected a consistent pattern, "quite unlike the skimpy evidence upon which the SST and aerosol debate was based." The record shown, collected at Mauna Loa in Hawaii by Scripps Institution of Oceanography chemist Dave Keeling, was initiated by the efforts of Roger Revelle, who was there testifying that day. Gore asked Pewitt, "Doesn't that lead you to look at it in a different light?"

One of the early deniers with no discernible expertise in climate science, Pewitt countered, "I am not an atmospheric scientist. . . I am a high energy physicist, a particle physicist. I understand false correlations; until one understands fundamental mechanisms of how things happen, they can be trapped into false correlations. Twenty-three years is not exactly a significant time frame in the weather."

"We have it going back to 1958 with reliability, but the same pattern goes all the way back to 1880, and you can see fluctuations for the Great Depression, for two world wars," Gore said. The chart clearly showed the dramatic drop in emissions during the Depression and the rapid increase following the period of industrialization after World War II.

Pewitt resorted to yet another of the Reagan Administration's favorite tactics. "You have several different effects going on here at the same time. It is a very difficult system to understand; this is a very complex physical system.... We have some bright people advising us on this program [and] nobody predicts anything to happen in less than 50 years. It is important not to waste the next decade, but it is also important not to stop everything in the world, on the basis of misinformation."

Gore parried, agreeing that we should avoid alarmism but "we also ought to avoid a head-in-the-sand attitude at the same time," before calling in another witness, Frederick Koomanoff, just appointed director of the carbon dioxide research program replacing Dave Slade. Dr. Koomanoff, who had only been in the job three weeks, had some experience as a climatic researcher and was not so quick to follow the administration's party line. He delivered a brief but solid call for moving ahead in a logical, prudent way, and as rapidly as possible. But Pewitt, interrupting, picked up his testimony almost where he had left off. He referred to a chart illustrating the correlation between fossil fuel burning and CO_2 buildup as very deceptive. "It is a clever piece of chartology, in that it is intellectually accurate, but can be subject to being read the wrong way."

The accusation of "chartology" is another item in the toolkit of the deniers. And as Gore pointed out, the longer-range chart, going back to 1880, showed virtually the same thing, and an even longer-range perspective appeared even more extreme.

I was beside myself, sitting in the first row behind the witness table, itching to jump up and demolish Pewitt's transparent polemics. It became increasingly clear to me that this wasn't about science but ideology. And the new administration's ideology was full throttle on "business as usual" energy policy and denial of environmental side effects—and even the cutoff of environmental research that might have some possibility of challenging the status quo philosophy of the Reagan "mandate," as they saw it. It was consistent with removing the solar collectors from the White House as a symbolic act.

Fortunately for Pewitt, the congressional committee had to adjourn at this point for a ten-minute vote in the House.

When the committee resumed the hearing, Gore invited me back to the panel for my reaction. I was glad I had the ten-minute break, too, since I could calm down and thus be more effective—especially since all of what is said is transcribed for the Congressional Record. What you say lives on—in fame or infamy, depending on how you conduct yourself.

Acknowledging that Pewitt was much more skeptical than the other expert witnesses on the panel, Gore agreed that it was "difficult to come to grips with this problem because you have such large areas of uncertainty. On the other hand, the consequences of the problem, as it is determined to be a real problem, are so enormous that it is not too early now to begin thinking about a response so we can eliminate the areas of uncertainty.. . . If our country and our society are to have a mature and intelligent response to this problem, then the tone of that response is critical. What do you think the proper tone of our response should be?" Gore and the other committee members looked at me expectantly.

I had a lot to say about the assessment of risks of climate change, but I tried to focus on the best way to get through to the policymakers on the committee.

Whether one is an alarmist or considers the CO_2 problem urgent isn't based on any scientific information. It is a value judgment. It depends upon how you personally fear potential risks versus how you personally fear the costs of mitigating them, versus your own political philosophy about whether individuals should be free to do what they want, or whether we have collective responsibility.

Given that, I am happy to give you my values, which are that I think the CO_2 problem is only urgent in the sense that a lot of its solutions are also related to solutions to other pressing issues, as I said earlier, such as developing crop strains, maintaining flexible energy supply options, [and not] changing soil horizons.

To me, these are urgent for other reasons than CO_2, and what CO_2 does is contribute to the need to maintain open options. In that sense it is urgent.

We are talking about incremental change, but we have to commit posterity to adapt to some level of CO_2. So, there is urgency in the sense we are already creating the commitment to some environmental change.

I do not believe that change will necessarily be catastrophic, although the issue of catastrophe is almost irrelevant. The issue is, if we have done something, and if we can know who did it, there is the question of equity as to who should be responsible or how we can minimize the damages so that it is not a catastrophe.

It is like many other pollution issues. I have a sense of urgency in the view that we need to consider now those actions which increase our options. In that sense, it is urgent. In the sense it will be an urgent catastrophe which will eliminate all life, that kind of statement is absolutely unwarranted.

Gore asked Pewitt if he wanted to respond. He declined. "His is a fair statement," he said.

Some of the names of the players have changed, and the level of uncertainties has narrowed in many—but not all—cases, but the same battle is still being waged over the future of planet Earth. I came away from that hearing with an unshakable thought: "I didn't know much about this guy Al Gore before today, but I am certain he'll be a major force in this field as long as he can stay in office."

Gore, of course, went on to change the world, but there is much more to tell about the fits and starts of that before this book is done. Pewitt and Edwards? They soon drifted into the obscurity they earned for themselves by their ideological disingenuousness—but not before delaying the process of climate protection research by many years.

Funds that were vital to scientific research and development were directed instead to military applications. As William D. Carey, Executive Director of AAAS, said of the Reagan DOE budget cuts in 1981, "We are letting the air out of our tires one pound at a time when we should make up our mind how much tire pressure will keep us out of the ditch."[5] In fact, the largest expenditures in energy research and development in the United States from 1975 to the start of the Obama Administration were in the Carter Administration, from 1977 to 1981.

In every congressional hearing from that one until the one in 2003 run by Senator John McCain, contention, denial, and overstatement became the rule, not the exception. We would have to wait until 2007, after the Democrats took over the chairs of the many environment and science committees in Congress, for civility from both sides of the aisle and serious witnesses to become the rule again. But in the decades between, indelible damage was done to our capacity to mitigate and adapt to climate change. Perhaps most serious of all was the development of deep partisanship over environmental policy. Concern for environmental issues, which began as bipartisan actions by such presidents as Theodore Roosevelt and Richard Nixon (the originator of the EPA), was transformed by the Reagan Administration to a conservative versus liberal issue. This is a complete misframing, since it is hard to find a more conservative word than conservation.

NUCLEAR WINTER WARS WITH CARL SAGAN

The attempts by scientists to discern the magnitude of the looming crisis soon received another setback—on an issue that had nothing to do with global warming. The Reagan Administration's insistence on rattling swords raised anew the question of what would happen in the aftermath of a nuclear war. Sadly, the search for answers would place me in conflict with one of the men I admired most. It would also hand conservatives additional ammunition to fire against all scientists.

In 1982, I received a preprint of a paper from Crutzen and Birks on nuclear winter, where they asserted that throwing smoke into the atmosphere would cool the planet. I could tell they hadn't calculated anything using climate models. We had a postdoctoral fellow named Curt Covey at NCAR, who had become interested in climate modeling, so I asked Curt if he would like to take a crack at it.

A little later we received a preprint from Carl Sagan, working with a group called TTAPS—Turco, Toon, Ackerman, Pollack, and Sagan. Rich Turco was at R&D Associates, in Marina del Rey, California. The research institute was classified "black," meaning largely secret. And Brian Toon and Jim Pollack and Tom Ackerman were all at NASA's Ames Research Center. They had devised a one-dimensional model to study volcanic eruptions and the effects of their dust veils on climate, and applied that model to smoke generated by nuclear war: It showed surface layer freezing after nuclear war. I wondered what would happen in a 3-D model and asked Curt if he would do that with me.

The effort involved modifying the radiative transfer package in the computer code written by others at NCAR, which was difficult. It so happened that Starley Thompson was a visitor at the time, working on his Ph.D. thesis for the University of Washington. He had said, "Don't bother me with any new ideas, I've got to finish my thesis work." Yet Curt, although an excellently trained planetary dynamist, wasn't a radiative transfer person. I knew that Starley had built radiative-convective models. Starley was the ideal person for the situation, but I had promised to leave him alone. So I did the only thing I could: Set out some bait.

As deputy head, I had an office on the fifth floor in the Advanced Study Program office area. A hall upstairs connected us, and that hall had long windows. Right in front of these windows ran heating ducts about belly-button height. Since we had about 20 pictures printed from microfilm to look at, in sequence like a movie, I laid them all out, all 20 pictures in a row on the longest surface available—the ducts in the hall—so Curt and I could analyze them. I laid them out right in front of Starley's office—by "accident," of course. Likewise, we were loudly talking about how the model was doing this and that. It worked—Starley couldn't stand it. He came out and said, "Wait a minute, how did you get this result and that result?" Soon we had him interested in the project, and he and Ramanathan helped Curt put in an aerosol routine. We then had a 3-D model that could actually begin to analyze the nuclear winter question, designed as a supplement to the work that was done in one dimension by TTAPS. Of course, Starley got proper credit for his work. It just took a nudge to get his attention—and he finished his thesis anyway, a few months late.

In the spring of 1983, Carl Sagan called a two-session meeting in Cambridge at the American Academy of Arts and Sciences (AAAS). The first meeting, which Carl chaired, was focused on the physical science; the second was on the impacts, or the biological science, and run by Paul Ehrlich. I attended both of them. Like good assessments that were done earlier, they recognized that it needed to be multidisciplinary and needed to include the "So what if the climate changes?" question.

The session reminded me of a time in Denver in 1977, at the AAAS meeting that Margaret Mead ran, when Carl and I had lunch. I remember saying to Carl that he was one of the best communicators anyone could ever imagine, but that he hadn't at that stage taken on the politically tough issues. I said, "Come on, Carl—we need you!" And he jumped in with both feet on nuclear winter.

The meeting in Cambridge that Carl had arranged was quite an important event. His wife Ann Druyan was probably the main influence in Carl's shift from being the wonderful communicator of science for its own sake toward science for policy and social relevance, because Annie was committed to arms control. I also think that Carl's

motivation was enhanced by Caspar Weinberger's loose lips about winnable nuclear war, which had reinvigorated the antinuclear movement.

The first few days we talked about science. The TTAPS results were
presented, and they were quite different from Crutzen and Birks's findings, because what the latter had examined were collateral fires—forest
fires associated with nuclear bursts. Rich Turco, who had access to classified data, knew what the Defense Nuclear Agency was interested in. It
wasn't trees—it was cities. The amount of fuels—tables, chairs, asphalt,
roofing shingles, building materials that burn in cities—dwarfed the
total numbers that were available in forests. Turco had calculated
how many millions of tons of soot aerosol would be injected into the
atmosphere. Toon and Pollack offered the one-dimensional radiative-
convective model they had used to study volcanic eruptions. They'd
had to modify that to add soot or absorbing aerosol, because the volcanic eruptions were generally considered to be reflecting aerosols, which
was Tom Ackerman's assignment as a radiative transfer scientist.

So how did Carl get into it, other than having been the adviser to Pollack and Toon? The answer is, he was a planetary astronomer who had
been observing the Mariner mission when it entered orbit around Mars,
and Mars was obscured. People wondered what was wrong with the resolution of the cameras. What they discovered instead was a raging dust
storm. Carl had recognized that what was happening on Mars was "self-
lofting." In other words, as the Martian climate moved toward summer,
the sun's rising heat generated strong winds, which then caused blowing
dust. But the dust, being large particles, absorbed enough energy that it
was actually heating the middle of the atmosphere, instead of the surface, thereby causing rising convection currents—that is, self-lofting.

Carl's contribution to the nuclear winter question was by analogy,
and this was a very important scientific component. He asked, "Could
we get self-lofting of soot, where the sun would then be absorbed not
at the surface [of the planet] as it is now, but will be absorbed in the

middle of the atmosphere in the soot, and therefore that would cause it to rise and continue to rise, and thus spread globally?" As in the Martian analogy, the soot would spread around the globe. In the 1-D model, the soot would settle in the upper troposphere and the stratosphere, and most of the sunlight could not get through—the result was nuclear winter.

TTAPS's model was running in what was global mean sunlight, roughly equivalent to the amount in March or September. It was not an extreme season, a winter or a summer. The Poles receive a tremendous amount of energy in the summer, and nothing the rest of the year; therefore, the annual average amount of energy is different from the amount that hits the Poles at any given date during the year. It was not, in a way, a good analogy to a three-dimensional world subject to the passage of time.

The reason Carl called this Cambridge meeting together was in the best tradition of good scientists: to invite people to try to shoot down the model. "Before we go public and make a big splash," Carl said, he would make sure the science was acceptable. He wanted to announce the nuclear winter effect in a meeting in Washington on Halloween, where there would be a downlink with Russian scientists—because after all it was the Russians and the Americans who were going to annihilate each other. Carl's concern was that we'd unknowingly created a doomsday machine, as in the movie *Dr. Strangelove.*

Carl also wanted to take a look at what the climate change would mean, and the scenarios were then handed to people in the second component of the meeting. Paul Ehrlich ran it, and it included the evolutionary biologist from Harvard, Stephen Jay Gould, and Robert May, who was then at Princeton. May, an Australian mathematical ecologist, later went to Oxford University and became the scientific adviser to two British prime ministers. John Holdren, an expert in both security and environmental issues, was there, too.

One night in Carl's hotel room, we decided that over the summer, all of us were going to do our best to reinforce scientifically the credibility of the TTAPS scenario, and if necessary point out flaws so that we—rather than our critics who were motivated to defend nuclear war fighting strategies—wouldn't try to damn our credibility later. I suggested that we make a pact—nobody goes public, nothing gets discussed until we meet in Washington. We would go to Washington two days early and make sure that we had a consistent story that was scientifically straight. Everybody agreed to this.

Curt, Starley, and I worked on the three-dimensional model all summer. Yet we immediately got a stunning result. When we put in an elevated soot aerosol and ran the model in perpetual January and perpetual July mode, we didn't come up with the perpetual deep freeze that TTAPS had. For one or two days after injecting smoke in July, temperatures on land did drop well below freezing. In other words, we initially thought we were validating the one-dimensional model result. But one day later, the very same spot that was well below freezing was well above freezing. Even worse were the variations. Coasts like California almost never went below freezing, except in the rare occasion when winds blew from the cold interior in the east. The 3-D model showed no uniform response.

The second drawback we discovered in one of the TTAPS runs. It determined the threshold where Carl said even a hundred-megaton war, which is a very small war using a fraction of the existing arsenals, was sufficient to trigger nuclear winter. Yet the run was flawed. That threshold at which enough smoke went in the air to drop the global mean temperature below freezing worked in the one-dimensional model, in a virtual world with perpetual, annual mean sunlight.

So we varied the amount of smoke and reached the conclusion that there was no such thing as a threshold war. In the real world, the plumes would drift around, and after the war, the combatants would have large blobs of smoke over them. Those blobs of smoke could, within a day, drop the temperatures well below freezing, as long as there weren't strong winds from the ocean to the land. Then the blobs would move away, or go out over the ocean, and the land would quickly warm back up. This was not a perpetual deep freeze—it was fluctuating freezes. So the threshold was an artifact of a one-dimensional model.

We discovered something else as well. When we set off the dust and smoke cloud in January, almost no cooling took place. At first I thought there was an error. The only place cooling took place was in the subtropics. And then it became obvious what was going on. There isn't very much sunlight in the winter in mid-to-high latitudes; therefore, depriving the winter hemisphere of sunlight didn't make much difference, because the temperatures were maintained in the interior of continents by very strong winds blowing over the warm oceans with their large heat capacities. That's precisely why Europe and the Gulf Stream are so much warmer than the Hudson Bay at the same latitude, or why San Francisco in winter is much warmer than Kansas at the same latitude. The smoke cloud made only a little bit of difference in January simulations over the U.S., Europe, and the U.S.S.R.

Starley looked at me and said, "Great, now we've told them when to push the buttons." We went through a soul crisis as to whether we should even discuss this in public. I remember thinking: But it's what we found. Not to discuss it is to play the same "ends justify the means" game of people whose ethics we didn't like.

I went to Walt Roberts and asked him what he thought. He said, "Are you sure you're right?" And I said, "We are not sure." He said, "Do some more experiments. Yes, you have to tell people, but you don't have to tell them right away. You can give yourself more time to be right."

He asked me to put the question in my own perspective. I said, "Well, it's essentially nuclear fall rather than nuclear winter, because of the fluctuating on and off. Still, nobody can grow crops in the fall."

But I wasn't going to use that phrase "nuclear fall" because I was media savvy enough to know that it would be used to attack Carl, conveniently ignoring the fact that a nuclear fall would be a disaster too.

I called Carl and told him what we were learning, and he did not really respond. A week or so before the Washington meeting, I got a copy of a *Parade* magazine article that Carl had written on nuclear winter and on why the nuclear arsenals had to be reduced by a factor of a hundred to drop below the threshold that would "trigger nuclear winter" and to eliminate the unwitting doomsday machine. Curt and Starley and I looked at each other and said, "But there is no threshold. It's a scientific fiction from a one-dimensional model."

I was in a bind. What was I going to do? I remember saying to my NCAR colleague Ralph Cicerone, "These guys have created a scientific extrapolation from 1-D, and we had an agreement to talk before we went public!" He advised me to "be straight, but don't make a big deal of it, given the remaining uncertainties" in all the components of the chain of logic. Good advice, which we followed.

When we met in Washington we had our advance session, as promised. Carl began the session by showing a video that he had produced with chilling electronic music, *The World of Nuclear Winter: The Deep Freeze*, and he even went so far as to say this could not rule out the extinction of humanity. I called up Starley and said, "Do you think this would lead to the extinction of humanity?" We both agreed—not nuclear fall. Not even nuclear winter, we thought, although it would be an unimaginable catastrophe.

I was more worried that the populations would be deprived of food, medicine, shelter, and national civic order than about the direct effects. The smoke plumes drifting over India could shut off the monsoon rains in the summer, killing more people by starvation than were living in the rich and combatant countries. But I couldn't imagine any scenario that included the extinction of humanity. Too many people lived in too many circumstances, either using relatively low technology or being too widely dispersed geographically to allow that to happen.

Again I was left with an ethical dilemma. The video had already been made, a *Parade* magazine feature was coming out, and Carl had written an article in *Foreign Affairs* as to why nuclear policy had to be revised, based upon the threshold concept I and my students believed no longer existed.

That began one of the most unpleasant chapters in my life. Starley, Curt, and I had talked—the credibility of science matters. Because we were studying global warming, any "ends justify the means" approach for nuclear war would come back to haunt us when we warned about global change years later. The credibility of climate modeling was at stake, we thought. So the question became how to finesse the problem.

I went to the Halloween meeting and, thinking of Ralph's advice, I explained matter-of-factly that there was no threshold and so forth. My low-key announcement was met with mostly stony silence. Some responded, "Well, these are very preliminary calculations. Vladimir Alexandrov and G. Stenchikov in Russia have done it using a GCM, and they get the same basic answer we do with TTAPS." That wasn't really true, though. In the first place, the Soviets were using a very low-resolution version of the 3-D Mintz-Arakawa GCM, and second, when I examined their results carefully, they were not getting the same answers as TTAPS. TTAPS was using mean annual sunlight, not July, like the Russians' simulation. We got the maximum effect in July too. We had a much higher resolution than Alexandrov, and Alexandrov used an extremely large scenario of smoke, much larger than we were using when we were identifying the absence of threshold. So I didn't think that the Russian study validated TTAPS or its threshold.

In private, I asked Carl to stay away from those iffy concepts so that we didn't have a public dispute that the media would use to distract from the fundamental message that we all agreed on, namely, that the idea of a winnable nuclear war is insanity.

At the public meeting, Carl put on a great performance. There was a downlink via satellite to Yuri Izrael, head of the State Committee for Hydrometeorology from the Soviet Union. This of course was viewed as disloyalty and cohabiting with the enemy by the war hawk, Richard Perle wing of the Defense Department. Donald Kennedy, then the president of Stanford, gave the introductory address. Walt Roberts supported the general notion, but had already been warned by me and was careful in saying that the details would change. But the basic idea would remain the same, which I genuinely believed, too.

When I got my ten minutes, I went over the general issues with climate models and drifting patches of dust. I emphasized the fact that even smaller wars than the Sagan threshold trigger could create fluctuating "quick freezes," but that it was hard to find a threshold in a 3-D model. I did not spell out that therefore the threshold was a chimera. I was hoping that Carl would do that for them, the same way I had ten years earlier for the "triggering an ice age" paper.

That's not what happened.

Months later, after giving a number of interviews on my views of the simulations, I got a call from Carl. He was very unhappy that some reporter had quoted me as saying there was no threshold. He kept saying, "What have you guys done wrong on your model?" And "You have to be careful of how you say this."

I said, "Carl, you have to back away from the 1-D model result. What you could say is, something got better and something got worse. What got better is that we don't have a trigger of nuclear winter. What got worse is that even smaller wars than your thousand-city hundred-megaton-war could create quick freezes. And in that fall-like condition, you can't grow crops."

We did not reach an agreement. I was supposed to change my findings, but I couldn't change them. The media was soon covering "Schneider versus Sagan," even though Carl and I were much closer in our philosophies than to Richard Perle. The *Detroit News* wrote up Carl as an exaggerator and me as the good guy. I had to write a letter back to them saying, I have very few fundamental disagreements with Carl Sagan; you have exaggerated this technical difference.

Some time later I got a transcript of a recent congressional hearing. I had testified to Congress many times, and I knew how congressional hearings work. You get five minutes, and then there's lots of Q & A. If you are very lucky, there's more than one or two senators or members there. Because this hearing concerned nuclear winter, and it featured Sagan and Richard Perle, it was well attended. One of the senators was the liberal Maine Republican Bill Cohen, who was later secretary of defense for Bill Clinton.

I read through the transcript, in which Senator Cohen asked Carl point-blank to explain the accusations of exaggeration, saying that work at NCAR and other places suggested that the effects were much smaller than Carl had said and that they were nuclear fall, not nuclear winter, and were unlikely to be as severe.

Carl replied, basically, that there was no difference between us. We got 20 degrees cooling, they got 20 degrees cooling, what's the difference?

I was frankly shocked. They derived 20 degrees cooling as a global average temperature decrease for a mean annual amount of sunlight. Our results showed 5 degrees under those conditions. Our 20 degrees were one- to two-day weather coolings in the middle of continents in the July case. I couldn't believe it. After all the conversations we'd had and my telling him that "the truth was bad enough," Carl was claiming that the threshold was still valid. It was clearly not scientifically tenable.

I walked into Walt Roberts's office, plunked the transcript down, and said, "Walt, what do we do about this?"

"Let me think about it—I'll talk to you in a few days."

Two days later, I got a phone call from the editor of *Foreign Affairs*. He said, "Walt Roberts spoke to me. I would like you and your colleague

Starley Thompson to write a *Foreign Affairs* article to follow up on the Sagan one, telling us what the science and the implications really are of this problem."

Starley and I had a long debate. We asked ourselves how much we wanted to be vilified. We decided that it was better to explain the issue than to continue to have the media distort it. Colorado Senator Tim Wirth, who held several excellent hearings I attended on this concern, told me, "You shouldn't let others define who you are or what you say, Steve"—referring to those framing me as the foil to Carl.

"How can I stop that, Tim?" He had no advice other than to try not to give them opportunities to dichotomize us. I sure tried—mostly in vain. In our *Foreign Affairs* piece we wrote that the extinction of humanity was a vanishingly low probability, but that killing a few billion people was enough to make anybody who had any ethics at all horrified. Even for a war in the winter, when we found relatively little additional weather effects, we asked readers to imagine surviving a winter with no deliveries of food, no heating, no electricity, no police or fire services, and no medical services. You don't need to know the weather forecast after a nuclear war to be deterred, unless you are already so insane that you can't be saved. And that's what we said in interviews and wrote in articles—over and over again.

We were criticized immediately by scientists who said *Foreign Affairs* wasn't a peer-reviewed journal; we should have submitted this to peer review. Indeed, we did publish all of it in peer-reviewed journals over the next year or two. We also explained it to our climate modeling friends at Livermore, Mike MacCracken and colleagues, and our other climate modeling friends in Los Alamos, Gary Glatzmaier, who had been at NCAR for a number of years, and Bob Malone. Malone was one of the developers of a third-generation general circulation model— known as a community climate model because it was in such wide-spread use. Six months later at Los Alamos they calculated the effects using a much higher resolution, much more "methodologically credible" model than we had. Bob called to say, "Hey, we get basically the answer you guys did, I bet you're happy about that."

I'm glad we were scientifically vindicated at the time, but that is the only positive that came out of this awful episode in my life. I remembered thinking to myself, "Working on this nuclear war problem does something to your head." How can some folks work on this stuff for their lives? Do they just make it abstract physics and forget the gigadeath implications? I couldn't wait to get out of this mess.

It had led to a schism between Carl and me, which was painful, because I admired Carl as a fine scientist and also as the single most influential popularizer of science. I had personally pushed him to get involved in controversial public issues back in the 1970s, but once he did—look at the thanks he got from me. I didn't wish to be an executioner of his credibility, and found the entire episode deeply painful from start to finish.

Carl wrote a critical response to ours in *Foreign Affairs*. Starley and I published a rejoinder, in which we were as polite and praiseworthy of their initial effort as possible, while defending what we had done scientifically. I was glad that a few years later Tim and Wren Wirth arranged for Carl and Annie and me to meet them for dinner in Washington, because it was a nice step toward reconciliation. That proved important, as we had to work together again to push President Bush senior to go to Rio for the UN environment conference in 1992. When Carl died of a blood cancer prematurely at the age of 62, I cried—though I'm not sure some who angrily remember our disputes would believe me.

One other piece of undeserved fallout came about from the TTAPS controversy—Ackerman, Pollack, and Toon were ordered by NASA Ames higher-ups to stop doing this work. Apparently, the Reagan Administration believed the work was hurting the Defense Department's nuclear war strategy. A handful of scientists were perceived as a threat to the DOD, and NASA's budget was implicitly threatened. The NASA chief who ordered them to stop acted on his own to protect his laboratory's budget from the risk of political retaliation. Many years later I met this man, who admitted he did it because of the climate around the Reagan Administration, not because he was ordered to do it. "How could you do such a thing?" I asked.

"I couldn't take the risk of losing the jobs of my staff and the important work that we do," he responded. It made me remember a passage from Dante's *Inferno* that was paraphrased by John F. Kennedy: "The hottest places in hell are reserved for those who, in a period of moral crisis, maintain their neutrality."[6]

With the fracas over a hypothetical war, climate scientists had made their first big splash in the media arena. I wish I could say that journalists and politicians became wiser as they became better acquainted with the issues. Yet the nuclear winter/fall flap was only prelude to the attacks that would come. Good science would soon achieve a victory, in helping to protect the ozone layer, but our efforts to remediate the wider problem of pollution had a much longer road to travel. The full fury of denial was upon us.

A FRAGILE PLANET GROWS ALARMED

4 **THE SCIENTIFIC COMMUNITY** would be exposed to the full glare of society at large in the late 1980s. Earth's increasing warmth began to attract public notice. Alarming news about the ozone layer led to banning the substances that damaged it. The terribly hot summer of 1988 in North America was followed by a new round of U.S. congressional and Canadian parliamentary hearings to investigate what was happening.

The Intergovernmental Panel on Climate Change (IPCC) was formed, and it would assess the seriousness of the situation that allowed governments to lay down guidelines they could follow to halt global warming. Nations thus started the process that would lead to the Rio Earth Summit in 1992, the first international accord on climate change. Yet these changes meant that many traditional corporate interests were going to lose money if controls on their emissions became law, and they wouldn't back down meekly. They responded by increasing their attacks. In order to win hearts and minds, climate scientists willing to enter the fray would have to join a rough-and-tumble free for all—no rules and no refs in this contact sport.

In 1985, the scientific community was taken aback when an article in *Nature* announced the discovery of the first large "hole" in the ozone layer, by British Antarctic Survey scientists Joseph Farman, Brian Gardiner, and Jonathan Shanklin. They had made ground observations of the hole in 1982 and spent more than three years collecting data before publishing their discovery. Scientists had known about ozone depletion since the mid-1970s, but the existence of a major area of ozone reduction in the polar region was truly alarming.

After F. Sherwood Rowland and Mario Molina, chemists at the University of California–Irvine, had published their hypothesis in 1974 that chlorofluorocarbons (CFCs) in the stratosphere were breaking down into chlorine molecules that destroyed some of the ozone layer, DuPont sanctioned a campaign against the science and scientists who announced concern for the ozone layer. They sponsored a one-month visit to the United States by a British denier of ozone depletion, Richard Scorer, who called the ozone-depletion theory "a science fiction tale . . . a load of rubbish . . . utter nonsense"[1] Dupont sponsored full-page attack ads in major newspapers citing Scorer and others to the effect it was all theory and no facts and misguided hysterical scientists behind a scare. But he and his cronies in the chemical manufacturing and aerosol industries were proven wrong. The battle to stop ozone depletion is one of the victories of atmospheric change prevention. It was fought not only in the United States but internationally as well.

The National Academy of Sciences had released a report in 1976 that verified the ozone-depletion theory, and it continued releasing reports as the years went on. The loss of ozone allows health-damaging ultraviolet (UV) solar radiation to reach the planet's surface in much higher amounts, leading to skin cancers in humans and adverse effects on agriculture and environmental systems. Molina and Rowland, with Paul Crutzen, were later awarded the 1995 Nobel Prize in chemistry for their pioneering work.[2]

Industry attacks in the late 1970s on the integrity and scientific acumen of these stellar scientists in front of Congress, in newspapers, and in other media by those only interested in maintaining market share was an object lesson of what we'd experience soon enough from the coal, oil, and automobile industries over global warming. Character assassination, out-of-context contrary facts, industry friendly Ph.D.'s denying that the science was "proven," and accusations that the concerned scientists were "doom and gloom" mongers trying to scare federal agencies into giving them research grants—all typical contrarian tactics—were honed into a fine political art in the ozone wars.

Luckily, governments were—eventually—more responsible. Taking the threat of ozone depletion seriously after the ozone hole was discovered, 20 nations met in 1985 to draw up a cooperative agreement for negotiating international regulations on ozone-depleting substances. The Vienna Convention for the Protection of the Ozone Layer took effect in 1988. It had no actual power to set reduction goals but supported an international treaty that set deadlines to phase out production of the substances believed to be responsible for ozone depletion, including CFCs. The Montreal Protocol on Substances That Deplete the Ozone Layer was first signed in September 1987 and went into force on January 1, 1989. All but one of the 196 member countries of the United Nations ratified the original protocol over the next 20 years, although not quite as many have ratified its continuing amendments. The international treaty successfully negotiated changes that resulted in proven decreases in atmospheric ozone depletion and the beginning of recovery.

Susan Solomon subsequently led an international team of scientists to the South Pole and directly measured the chemical composition of the polar stratosphere in order to provide a clear smoking gun of human pollution as the prime cause of the ozone hole. Her team's in-the-field press conference announcing this finding noted that not only had they found the "smoking gun," but they also knew who pulled the trigger and at whose head it was aimed.

After this establishment of "proof" that humans caused the ozone hole, the Montreal Protocol, which banned about half the emissions of ozone-depleting substances, was expanded to cover about 90 percent of those substances. Companies were ordered to find replacements for CFCs and other substances to limit the damage to the atmosphere. Who was in the best position to undertake such research and development—and ultimately to patent and profit from the market for these new ozone-safe products? The same folks who were producing the CFCs, aerosol cans, and so on—American chemical industry giants like DuPont. They promptly switched to the production of hydrochlorofluorocarbons (HCFCs). To its credit, the chemical industry has not abandoned its new green stripe, even to this day, on this problem—an undeniable achievement for which they should be praised. That "it's not too late to learn" is the message I hope the fossil fuel industry will take from the ozone-chemical industry brouhaha.

An additional note on this accord is relevant to a climate change treaty. Some nations, including India, China, and Russia, were reluctant to join in the Montreal Protocol because they had just geared up to manufacture CFCs, and now the rich countries had invented a new trick to keep them out of the market: ozone depletion. Moreover, the developed, technologically advanced countries had conveniently gotten a decade's head start inventing substitute chemicals and industrial processes that could adapt to the new products. So the treaty recognized this equity issue and set up special financial instruments to compensate them for losses of potential CFC sales—and then they too joined.

The relevance to climate treaties yet to come is that we have to recognize that different countries are in different development positions when the treaty is required. Complex and individualized side payments or other forms of compensations may be needed to get everyone on board.

"What would it take to declare a 'climate hole'?" I was asked a few years later by Paul Crutzen. He was of course saying that it would be a tough sell to claim any one extreme weather event is "proof" of dangerous climate change.

Not surprisingly, HCFCs and hydrofluorocarbons (HFCs) were later discovered to be measurable contributors to anthropogenic global warming. No change could be expected under George W. Bush, but at the end of the first hundred days of President Obama's term, his administration announced that it was considering asking the 195 nations who had ratified the original Montreal Protocol to amend the treaty to require mandatory reductions on HFCs. If the amendment were taken up, it would be the first time ever that the United States has proposed a mandatory cut in greenhouse gases internationally. HFCs can warm the climate up to 10,000 times more per molecule than carbon dioxide and do the most damage during their first 30 years in the atmosphere. There are many more CO_2 molecules, so it is still by far the principal human-induced greenhouse gas, but taken together, the others constitute almost half of the heat trapping and can't be ignored. Eliminating HFCs now has a good chance of making a significant difference before 2050.

HEAT BECOMES HOT NEWS

In May, June, and July 1988, unprecedented heat waves gripped the eastern two-thirds of the United States and parts of Canada, and the mainstream media rediscovered global warming. Essentially the problem escalated from the reports of a few hundred climate intellectuals, a score of science writers, and a handful of politicians to the front covers of print media and top of the television news stories—even to the political establishment, which now had to decide whether or not to do something tangible about greenhouse gas emissions.

The intense heat baked the central and eastern parts of the continent. Drought losses to crops and other related industries in the United

States alone amounted to more than $40 billion over the summer. Even more costly, according to the National Oceanic and Atmospheric Administration (NOAA), was the loss of human life—an estimated 5,000 to 10,000 people died of causes related to heat stress. A similar heat wave in 1980 had caused half that amount of monetary damage but just as many deaths. The very young and the very old or chronically ill were most susceptible to the physical stress of high temperatures.

The media came out to NCAR all the time. Soon the NCAR press service was going crazy. Warren Washington and I and other scientists appeared in probably 50 video clips a month on this problem.

Senator Tim Wirth of Colorado brought a number of experts to Washington to testify before the Energy and Natural Resources Committee, first and most famously Jim Hansen.

"The Earth is warmer in 1988 than at any time in the history of instrumental measurements," Jim told the assembled senators, who must have been sweating in the record 98-degree heat on June 23. "The global warming now is large enough that we can ascribe with a high degree of confidence a cause-and-effect relationship to the greenhouse effect. . . .. Our computer climate simulations indicate that the greenhouse effect is already large enough to begin to effect the probability of extreme events such as summer heat waves . . . and it is changing our climate now." Jim told reporters, "It's time to stop waffling so much and say that the evidence is pretty strong that the greenhouse effect is here." Jim also said he was 99 percent sure that the hottest year was not a natural fluctuation, which strictly speaking was not precise, but certainly was well over the threshold for concern.[3]

The next day the *New York Times* declared "global warming has begun." More congressional hearings were held by Gore and Bradley and Wirth, in which a number of scientists participated, including me.

Robert Watson (later to chair the IPCC) and I briefed Senator Bill Bradley, a democrat from New Jersey, before the Energy and Natural Resources Committee in August and September.[4]

In my submitted written testimony, I commented, "Quite simply, the faster the climate is forced to change, the more likely there will be unexpected surprises lurking."

Senator Bradley fixed on that little word, "surprises."

"I presume that you can tell me what some of those surprises are that might be lurking." He went on that he would like to ask the experts to "address the questions that you pose in your testimony, which are what can happen when we talk about global warming. What can happen? What are the probabilities of alternative outcomes that you foresee? And how did you determine these odds? And what are the potential consequences should the outcomes occur? What are the surprises lurking?"

"A surprise I suppose is by definition something that you don't know what is going to happen," I replied. "Dr. Watson just said that we know that there have been rapid changes in the past that have occurred in the climate in ancient times that are not understood." One example of a surprise, I told him, occurred about 11,000 years ago, when the world was about finished with the Ice Age, and then in a matter of less than 100 years, it was plunged back into almost Ice Age conditions again. And this period lasted something on the order of 500 years, eventually followed by our present warm interglacial period.

"That is an issue that is now getting considerable attention because it occurred relatively rapidly and most of our climatic theories would suggest that the climate would change more slowly than that kind of change," I explained. "There may be surprises of a different nature rather than just quick glaciations." I continued, "This is speculation, of course. The Arctic Ocean is covered with sea ice now. And if one warms the climate to a certain point, it is possible that the nonlinear processes that Dr. Watson mentioned could have that ice all of a sudden disappear if one passes a certain threshold of warmth which would then have implications for atmospheric circulation patterns that we haven't begun to explore."

Bradley requested me to quantify the threshold. "How much would the temperature have to go up before the glacial ice would be melted?"

"I certainly can't quantify it now," I replied. I explained that we were "trying to look at how much energy it would take. It is not the same thing as how much temperature it would take."

Bradley wouldn't give up. He wanted numbers.

"Well, can you quantify it not specifically but in orders of magnitude? We have heard that the temperature of the world is one degree higher this year. How much more would it have to be over time?. . . Five degrees, ten degrees?"

I did the best I could to give a responsible answer, in Fahrenheit, as I recall. "Five probably. The U.S. Geological Survey is starting to look at a period in Earth history when we believe there was no ice in the Arctic Ocean area, at least certainly not in the summer. And that is called the Pliocene, something like three million to four million years ago. And what they are trying to do is find out exactly how warm it was then, what was the sea level then, and is this a good metaphor. In a sense, has Earth performed this experiment for us?"

"But what can you tell us today?"

"Five degrees."

Senator Bradley continued to make his point. "I only say this because if you try to explain this issue in terms of references to 10,000 years ago and debates within the scientific community, the urgency that all of you feel and that many of us feel is diminished. So, in posing your questions to you, this is your opportunity to make this issue more accessible to greater numbers of people.

"So, what can happen? What are the probabilities of alternative outcomes? And how did you determine the odds? And what are the potential consequences?"

I gave my usual measured responses, despite embedding serious concerns of our present course and warnings to be precautionary. I knew that as a result of those committee hearings, people were talking about actual legislation to regulate emissions—meaning corporate interests were being threatened. If they could paint me as wild-haired, I'd do their work for them.

That point was driven home during a major international meeting in 1988, the Toronto Conference on the Changing Atmosphere, which was set up to bring environmental groups, media, governments, and scientists together to discuss action on climate change. The Toronto Conference was run—for the first time—not by the scientists and government establishments, but by nongovernmental organizations (NGOs). They had an agenda: to advocate a protocol or a treaty calling for cuts of 20 percent in the CO_2 emissions of the countries of the world by 2005, in particular the rich countries. This was a radical agenda, because it involved measures that would actually cost various interests significant money. The meeting convened in June 1988, attended by scientists and policymakers from 46 countries and organizations—and plenty of reporters. Climate change at last had caught the attention of the mainstream media, not just the dozen usual science writers.

I was placed in the unfamiliar position of being a conservative among all the environmental NGOs. I was very concerned that nobody had analyzed carefully what a 20 percent cut by 2005 meant. How much would it cost? Who would win, who would lose, who would pay? How fast could we accomplish this goal even if the political will could be found? I raised all these concerns, to the consternation of some of the radical groups.

"Suppose we announce that all nations have this target of 20 percent in emissions cuts to begin to be implemented in five years—unless nations can show why it is not feasible or desirable or have other alternatives to make the world safer climatically?" That was the line I took—we needed to do due diligence before we just announced numerical targets. My position reminded me of nuclear winter becoming nuclear fall thanks to my insistence that we can't just assert outcomes. We have to do the relevant calculations first to see what such a cut would imply for both environment and economy. Even back then I was separating myself from some in the environmental movement who seemed to believe in numerology—hard numerical targets that were politically charged but were not well grounded scientifically. I preferred framing it as risk management: We'd be nuts to take such big risks without trying to slow down climate change, but we don't know precisely what levels of warming or emission were safe or dangerous and so could not justify a specific numerical target.

My second note of caution at Toronto—and in editorials in *Climatic Change* and at press conferences—was my interpretation of the heat wave: Nature, not humans, makes heat waves. New Zealander Kevin Trenberth, who had come to NCAR in 1984, had already showed that the most important reason for the intense heat was a supersized La Niña—the opposite to an El Niño—in the Pacific Ocean. In fact, Kevin believed in anthropogenic global warming; he was just trying to show that year-to-year fluctuations were mostly natural. For his scruples, he found the media using him as a contrarian. "I believe in global warming and because I challenged that it caused a heat wave they say I don't believe in it," he lamented in my office one day in June 1988. "I now understand what you have been going through." I deeply appreciated the camaraderie that emerged in our mutual frustration in being misframed out there.

As for myself, I received some nasty e-mails, not just from the usual deniers, but from the more radical environmental groups, because my statements about natural variability got media attention. "You can't take this heat wave away from us," one said.

"I don't want to," I replied, "and you can use it to show how vulnerable we are to such events, which will become increasingly commonplace as global warming takes off." I warned that when the weather turned unusually cool soon thereafter, critics were certain to try to damn their credibility.

I later wrote a book called *Global Warming: Are We Entering the Greenhouse Century?* in which I described what had happened in the 1988 heat waves, and how frustrating it was after having tried for 15 years to focus people's attention on this problem, and then a random fluctuation (the prolonged and deadly heat wave) comes along and, even though global warming added a bit to the severity, the problem becomes exaggerated.

I found myself in the ironic position of writing,

No, ladies and gentlemen, this is not global warming, because the world has only warmed up about a half degree Celsius, and this heat wave is many degrees warmer than normal. We've only been responsible for maybe ten percent of it. However, if you don't like this, stick around, because this is the kind of event that we are going to have increasingly frequently in the future. Moreover, random or not, it most definitely exposes our vulnerability to an event that will become increasingly frequent as we warm up. And therefore it is symbolically important evidence of the reason we need to move forward with policy to slow it down.

I think we have to play this one straight. The truth is bad enough.

Another point I made in *Global Warming* came in a chapter I called "Mediarology," about how difficult it is to communicate honestly complex science through the media. I told stories about the distortions that go on—how the media, trained in political reporting, generally has the view that if you get the Democratic opinion, you must get the Republican. You give equal time to all claimants. I argued that this "doctrine of balance" is pernicious when applied to science, because science is rarely just two-sided like two-party politics, where balance is appropriate. Scientists winnow out the relative likelihoods of all of the various potential outcomes. We are not in the business of equal time for all claims; we are in the business of quality of evidence assessment. Therefore, what we need to do is report the relative strength of the arguments, not give equal time to all claimants of truth.

Truth is not precisely known, and sometimes not even fully knowable. "Truth" is approximated by increasing refinements via a series of subjective probabilities and possibilities, updated by new data and theories assessed by a community of those informed on the issues—because that's all science can offer about the future.

The denial apparatus was not very organized in 1988. They didn't need to be, because they weren't feeling threatened by a bunch of climate intellectuals. But once Congress and other legislatures around the world started talking about adopting real policy measures, the industry got organized at breakneck speed.

Scientists like climatologist Patrick J. Michaels and environmental science professor S. Fred Singer, both at the University of Virginia, and Dick Lindzen, who had long been the opponents of science for policy, jumped in with their contrarian views. They were counting on "media balance" to give them equal status in the debate, no matter how small their numbers relative to mainstream climate scientists. They were essentially handed a huge megaphone by the fossil fuel industry and its allies, and by the ideologues who did not believe in protecting the commons if it hurt return on private investment. That's when the ugliness really took off. Big oil and big auto, joined by several banks, and even the U.S. Chamber of Commerce, formed the Global Climate Coalition (GCC)—ostensibly to give a balanced account unlike, in their framing of the issue, the biased hysteria they accused the environmental groups of promoting—and most mainstream climate scientists too, by implication.

One of the tricks used by the GCC and the industrial lobby against climate policy was character assassination. They followed what the chemical industry had done a decade earlier in the 1970s, when they attacked Cicerone, McElroy, Rowland, and Molina over the ozone depletion area. Soon GCC targeted scientists such as Hansen, me, and others who were discussing the issues as a public policy concern. They testified before Congress, entertained the media—and sowed nasty contention about this problem, as it switched from an intellectual discussion to one of whether strictures should actually be enforced.

Several years before the heat waves, I had squared off against Pat Michaels at a fairly cordial, but sparsely attended, hearing called by Indiana Democratic Congressman Phil Sharp. Almost nobody in Congress came to witness it, since climate change during the Reagan years was a nonstarter. I recall one of our exchanges when I advocated a carbon price—via a tax, perhaps. "Are you kidding me," Phil said, "in this administration? You're joking!"

I said to him afterward, "In not too many years, Congressman, when events change perceptions, we can point back to our hearing and, for what it is worth, tell them, 'I told you so.' "

The aftermath of the heat waves brought about a remarkable difference. When Phil held another hearing, it was jam-packed, with cameras everywhere. Many members of Congress attended to make statements and ask questions. When I went up for my turn at the microphone, I reminded him of our earlier hearing and said something in jest that now is the time for us to say "I told you so!"

The disinformation campaign was in full swing. Jim Hansen, in a later address to the National Press Club commemorating the 20th anniversary of his "stop waffling" testimony to Congress, described the tactics of the fossil fuel industry. "Special interests," he said, "have blocked transition to our renewable energy future. Instead of moving heavily into renewable energies, fossil companies choose to spread doubt about global warming, as tobacco companies discredited the smoking-cancer link. Methods are sophisticated, including funding to help shape school textbook discussions of global warming. CEOs of fossil energy companies know what they are doing and are aware of long-term consequences of business as usual. In my opinion, these CEOs should be tried for high crimes against humanity and nature."[5] Jim was never a man to mince words.

Journalist Ross Gelbspan wrote a book called *The Heat Is On* detailing his investigations on the conspiratorial nature of the GCC strategy. He called them and the scientists and critics they supplied with contrarian arguments as "interchangeable hood ornaments on a high powered engine of disinformation" paid for by industry. Citing leaked internal GCC documents, he reported that their plan was to "reposition" the debate as "theory, not fact," creating widespread doubt and uncertainty, by pushing the minority views of contrarians like Lindzen, Michaels, Fred Singer, and the like. This was vintage Tobacco Institute strategy, and once again, it worked. As Gelbspan put it: "By keeping the discussion focused on whether there is a problem in the first place, they have effectively silenced the debate over what to do about it."[6]

One technique they used in the 1990s was aptly described by blogger Aaron Swartz: "The media, in response to flurries of 'blast faxes' (a technique in which a press release is simultaneously faxed to thousands of journalists) and accusations of left-wing bias, began backing off from the scientific evidence. A recent study [in 2004] found only 35% of newspaper stories on global warming accurately described the scientific consensus, with the majority implying that climate scientists who believed in global warming were just as common as global warming deniers (of which there were only a tiny handful, almost all of whom had received funding from energy companies or associated groups)."[7]

Sociologist Riley Dunlap and Aaron McCright later did an analysis of the number of times the "elite six" climate scientists testified to Congress versus nonexperts.[8] Before the GCC campaign, we—I was one of the "elite six" along with Jim Hansen, Sherwood Rowland, and others—were the experts, testifying about four to one relative to nonexperts. By the early 1990s the ratio changed; now, half of the time, at most, we were the witnesses. The rest of the time, people with little or no expertise—but who were ideologically in line with those who denied climate change—got equal status at the table despite their virtual lack of knowledge. Not surprisingly, the momentum for climate policy in 1988 was totally lost by 1992. In 1994, with the victory of the Republicans in the House of Representatives, climate policy was dead in the water, even though Al Gore now occupied the Vice President's chair.

In a further analogy to the tobacco lobby, it later turned out the GCC knew their strategy was potentially dangerous and initially withheld the information from their industry defense documents. Andrew Revkin in his *New York Times* blog, Dot Earth, recently—14 years after the fact—obtained documents proving that the GCC's own 1995 internal science advisory committee had told them their strategy was not scientifically viable. Revkin reported that the coalition removed a section from the committee's internal report that said " 'contrarian' theories of why global temperatures appeared to be rising 'do not offer convincing arguments against the conventional model of greenhouse gas emission-induced climate change.' "[9] It did not appear in their public documents for three years. An amended version of the "backgrounder" was eventually published in 1998, "but the coalition continued to question the scientific evidence that greenhouse gas emissions could heat the planet enough to justify sharp cuts in emissions." The GCC never put out in public the warning from their own scientists that climate change was a "potential threat."

The timing here is interesting. The GCC withheld the scientific viability argument until 1998—shortly after British Petroleum CEO Sir John Browne came to Stanford University and gave a widely publicized public talk in which he said that BP was withdrawing from the GCC. It was changing its name to Beyond Petroleum and would enter the renewable energy business, because climate change was a real threat. Soon thereafter other major companies withdrew, and the GCC eventually collapsed. But a decade of doubt and delay was its legacy—exactly what it was set up to accomplish.

THE BEGINNINGS OF THE INTERGOVERNMENTAL PANEL ON CLIMATE CHANGE

Despite the confusion in the public mind, the scientific community was well aware of the dangers humankind was courting. Bert Bolin, respected professor of meteorology at Stockholm University—and my lone supporter at the 1974 Stockholm meeting—was asked by the UN Environment Programme and the World Meteorological Organization to convene an international assessment on climate change. In 1988, I ran into Bert in Washington at one of many National Academy of Sciences studies. My climate science colleagues at the academy and I had once again reaffirmed that while nobody could be sure, the warming from CO_2 doubling was likely to increase 1.5 to 4.5 degrees Celsius (2.7 to 8.1 degrees Fahrenheit).

He told me about the possibility of creating what would become the Intergovernmental Panel on Climate Change, and he asked me what I thought.

"Bert, I think it's a terrible idea."

Surprised, he said, "Why is that?"

George H. W. Bush was now Vice President and would likely be running the country after 1988, and his administration would be strenuously arguing that until we had more information, we should have no policy. I used to call him the "Climate Monkey-in-Chief"— see no climate change, hear no climate change, and speak no climate change. I think his son even more deserved that title years later.

"This is just going to be another two years to do yet another assessment. It's going to provide another excuse for them to call for delay. And what are they going to learn that we don't already know from assessments made in Australia and the U.K.? And we've issued at least five or six assessments by now, right here from the National Research Council."

Bert had another view. "But how many of those are convincing to people in India or Indonesia or developing countries who don't trust the science that comes out of it?"

"Yeah, Bert, I guess you're right," I admitted. "But we have had so many assessments and the scientists are all so overworked doing assessments, we can't get enough time to do our own research to help answer the many remaining uncertainties."

Bert said, "You're right on that. But what happens if we don't have an international consensus? Will it be possible to have climate policy without having a scientific group in which various countries in the world have some political ownership?"

I thought about that point. "You're right, I guess it has to happen," I agreed. "You've convinced me."

Over the next two years the IPCC met, and the combined effort was phenomenally successful. The whole story is told beautifully by Shardul Agrawala, then a graduate student at the Woodrow Wilson School of Public and International Affairs at Princeton, in an article in *Climatic Change*.[10] My role in the establishing of IPCC was minimal, since lead authors were nominated by national governments and my well-publicized critiques of the Bush Administration policy did not place me at the top of their list. I was, however, sent several draft chapters to review by the IPCC Working Group I, and I made extensive reviews and suggested new language. They used some of my suggestions, and when the Assessment Report was published a year later, I was listed as a contributing author. It was flattering they thought to acknowledge me, since I spent only a dozen or so hours on it. That was a luxury I would never be so fortunate to experience again, as the effort involved in the successive three IPCC assessments was so time-consuming, I used to joke at speeches that IPCC was my "pro bono day job."

The first IPCC report in late draft stage provided the primary basis for scientific credibility during the Second World Climate Conference in Geneva in the fall of 1990. When I went to that conference, I was stunned at its difference from the first world conference. First of all, Sir John Mason lost his argument that it would be premature and irresponsible to have political leadership involved in the climate issue until we had resolved to a very high degree of certainty all the scientific issues. This meeting was dominated by political leaders. Unfortunately, the only really good science there was supplied by the IPCC. By this time both the environmental advocates and the industrial ones were already in an "end of the world" and "good for you" false dichotomy debate. The battle lines of contrarians versus deep ecology were drawn.

The Second World Climate Conference was divided into two parts. The first featured a series of nongovernmental scientific and technical sessions that resulted in a strong statement about the risks of climate change. The second part was a discussion among heads of governments and ministers from 137 states and the European Community (this was before the EU). "Discussion" hardly begins to describe the hard bargaining and differences in opinion among these delegates. The resulting declaration did not specify emissions reduction targets, which angered some, but it did recommend pursuing a framework treaty on climate change, with negotiated commitments and solutions, in time for adoption by the UN Conference on Environment and Development in June 1992. That world conference, which took place in Rio de Janeiro, is better known as the Earth Summit.

MOVING FORWARD

At the same time that the stakes were being greatly raised in the world at large, my own corner of it was undergoing change as well. As much as I had loved my term at NCAR, their management's response to a budget cut was to protect its traditional components—and backed away from building the new global change research. When I reminded my bosses that they had promised an expansion in our area, they said, "At least you didn't get a cut like others." But not all others were cut, and it was clear that NCAR's priorities did not include creating the new interdisciplinary climate systems division that Starley and I were advocating. I was told point-blank it was "too radical."

Around that time Paul Ehrlich called me and said that a slot was opening up at Stanford University soon in the Department of Biological Sciences and the ecology and evolution subgroup, which consisted of scientists whom I had known and worked with for a long time, particularly over the nuclear winter problem. I knew them to be brilliant ecologists, interested in large-scale problems. They said to me, "We don't have a climate person or a globalist. Our students are being deprived of being connected to the largest component of the biosphere, namely, the climate system." Even though I was also considering an opportunity to create an environmental institute at the University of Michigan (which had the additional lure of being close to my future wife and then current research partner, Terry Root, who was a professor there), I decided to go to Stanford for complicated reasons both personal and professional. Stanford had just created an Earth Systems Program; I taught in the very first course in 1992, and I've been there ever since.

ON THE ROAD TO RIO

The Earth Summit in Rio de Janeiro was the "real" meeting of nations, from which the UN Framework Convention on Climate Change (UNFCCC) emerged. The phrase on everyone's lips was also the objective of the convention: "stabilization of greenhouse gas concentrations in the atmosphere at a level that would prevent dangerous anthropogenic interference with the climate system." We are still to this very day arguing about what that means, because the UNFCCC did not define it. And it's clear that what's "dangerous" is not a scientific judgment, but rather a value judgment about what's important and what is acceptable or unacceptable risk, and to whom. Plus, to make that value judgment, we have to know how much climate change will take place. And we have to know what will be impacted, what will happen to wildlife, to fisheries, to coastlines, to human health, to agriculture, to forests, and so forth. And how do we weigh a change in agriculture in one country versus a threat to health in another? What could be dangerous to one country could be a benefit to another. None of that was defined. However, this objective determined the agenda to occupy IPCC in the 1990s and to this day.

The UNFCCC set up the Conference of the Parties (COP) process, which then assumed authority throughout the 1990s to hold a series of meetings of national governments in which a climate protocol was to be hammered out. The meetings were highly contentious. The problems were accentuated by the presence of a large number of media, by NGO communities, and by the independent scientists and academic organizations who registered as nondelegates but sure had a lot to say. All of this interaction derived from the Second World Climate Conference's push for a "real" meeting.

The United States came very close to denigrating the Rio Earth Summit. Right up until the meeting, George Bush senior had not decided whether to attend, despite the fact that most world heads of state planned to go. I was invited to a meeting in Washington in the spring of 1992 to try to help persuade the President to attend— though not directly, of course. Carl Sagan and Al Gore, by then a senator, had assembled the leadership of the major religions in the United States. The premise of the meeting was that although we might not agree about the origins of life, we all agreed about the need for stewardship of the planet. Climate change was one of the prime issues, and we wrote a joint statement about how the United States must be a key player in continuing our world leadership, which we had been losing since the Reagan Administration's effective denial of global and environmental problems.

This "get the president to go to Rio" group was a coalition as unusual as it was powerful, bringing well-known scientists like Carl Sagan and Ed Wilson and Stephen Jay Gould of Harvard together with leaders from Baptist and other Protestant denominations, the Catholic Church, the Jewish faith, and Islam. At one point Steve Gould, who had been fighting a long battle with these religious leaders about the teaching of creationism in schools as an alternative to scientific evolution, came to the microphone. He looked at the scientists and said, pretty much, "How can you be talking to these people? These are the very same people who are trying to rewrite the nature of science, do not understand that science and religion are different, and who have been leading pernicious campaigns against intellectual values."

It almost brought the house down. Carl Sagan rose and gave an impassioned speech, saying that we are not here to argue origins—we are here to argue common interests.

Al Gore then took over the microphone. "I am a man of deep religious faith. Yet I believe in stewardship. I believe that God's requirement of us is an interpretation of our stewardship responsibility. I am thrilled that the bulk of scientists have goodwill and religionists can put aside their differences about origins and move forward in protecting the planet."

Gore really saved the meeting. The next day I was invited by Al and Carl to be a last-minute replacement for Steve Gould in lobbying members of Congress and the Senate. I welcomed the opportunity, because I was able to reconcile with Carl for the first time since the nuclear winter era. We were on the same side, and it was fun fighting together again.

The representative of the United States Conference of Catholic Bishops said to George Mitchell and Bob Dole, the Senate majority and minority leaders respectively, that they would be reading a sermon supporting stewardship the following Sunday, and "there are more people in those churches than there are people in our schools. I think you would do well to listen to us, and I think you would do well to inform your president that if he continues to deny his participation in this international meeting, he will find that the church groups will not consider that to be an appropriate ethical behavior."

Shortly afterward President Bush announced that he was going to Rio. And he did indeed sign the treaty, which committed the United States to voluntarily cut back its greenhouse gases in the late 1990s to near its 1990 level. Of course by the end of the decade, we were up by 15 percent, not down by 5 percent.

I thought a further positive note was struck later in 1992 with the election of Clinton and Gore, ending a 12-year period of frustration over lack of climate policy. Terry and I even attended the Environmental Inaugural Ball in Washington in early 1993. The future was finally looking up, I thought. Perhaps I should have been less naïve.

THE BATTLE HEATS UP AND SO DOES THE WORLD

5 ::::: **TWELVE-YEAR-OLD** Severn Cullis-Suzuki traveled 5,000 miles from her home in Vancouver, Canada, to represent the Environmental Children's Organization (ECO) at the Earth Summit in 1992. She and three friends had raised the funds to pay for their trip to Rio de Janeiro, because they felt that someone of their generation should speak to the adults—the decision-makers—at the conference about their fears and their hopes for the planet.

Severn, who founded ECO, is the daughter of geneticist David Suzuki, a well-known Canadian science broadcaster and environmental activist. She had a real impact on the Rio delegates with her heartfelt appeal. From the podium she spoke to the assembled science experts and government officials without quavering. "I am fighting for my future . . . I'm here to speak for all generations to come." She looked them straight in the eye and raised her voice in her plea for Earth. "If you don't know how to fix it, please stop breaking it!"

The video of Severn's 1992 speech has been viewed many thousands of times on YouTube, where it is entitled, "The Girl Who Silenced the World for 5 Minutes."[1] Comments are posted daily, some by viewers who don't even realize that her speech was given more than 15 years ago. That's because Severn's message is just as relevant today. She has continued her social and environmental activism, encouraging Canada's youth to speak out for their future and pursuing graduate studies in ethno-ecology. We need more courageous young people like Severn who grow up to become leaders in the climate science battles.

These battles intensified following the Rio Summit. Even though the UN Framework Convention on Climate Change (UNFCCC) was signed at Rio by 154 states and the European Community, not everybody was happy about the idea of committing to global action. In particular, President Bush senior found the idea of the United States taking orders on climate change strategies from any other entity—whether the United Nations or another global committee—to be anathema. He regarded the U.S. environmentalists' efforts to join with other countries in establishing limits on greenhouse gas emissions as a threat to national security. At Rio he remarked, "The American way of life is not up for negotiation."

A second prong of the Bush argument against global treaty participation emerged from the President's close ties to the oil industry. Imposing outside restrictions on the U.S. fossil fuel industry was not an option they would agree to. Nevertheless, despite the ideological rhetoric, Bush did the right thing for history: he signed the UN Framework Convention on Climate Change that called on all signatory nations to stabilize "greenhouse gas concentrations in the atmosphere at a level that would prevent dangerous anthropogenic interference with the climate system. Such a level should be achieved within a time-frame sufficient to allow ecosystems to adapt naturally to climate change, to ensure that food production is not threatened and to enable economic development to proceed in a sustainable manner." Moreover, the U.S. Senate ratified the treaty, making it the law of the land.

THE POWER OF A WORD

In December 1995 the IPCC finalized its Second Assessment Report after months of difficult bargaining over the final approved document.

Published in June 1996 as *Climate Change 1995*, the Working Group I SAR concluded that "the balance of evidence suggests that there is a discernible human influence on global climate." That word "discernible" remained in the report only after an ugly battle in the plenary session in Madrid in November. Some 2,000 scientists and experts from many countries around the world participated in the IPCC working groups, reviews, and discussions, but in the end, the main obstacle came down to several countries whose representatives seemed intractable.

Since the first IPCC Assessment Report in 1992, the preparation of all IPCC reports and publications has followed strict procedures. The work is guided by the IPCC chair and the Working Group and Task Force co-chairs. Hundreds of experts from all over the world participate in the preparation of IPCC reports as authors and contributors, and many hundreds more as reviewers. The composition of author teams reflects a range of views, expertise, and geographical representation. Review by governments and experts are essential elements of the preparation of IPCC reports. The procedures were developed further over the years. The chart below, from the IPCC website, offers a clear view of the process.

The Second Assessment Report (SAR) had three working groups, each devoted to a special area of interest, a multidisciplinary effort intended to form a complete picture of climate change during the period of time surveyed. (That same allocation of working groups continues to be used for future reports.) In the SAR, Working Group I was headed by atmospheric physicist Sir John T. Houghton and Brazil's L. G. Meira Filho, and its section of the report was called "The Science of Climate Change." I served as lead author of Chapter 11, on advancing our understanding of the science. Working Group II was co-chaired by my old friend Bob Watson and M. C. Zinyowera, and its section was called "Scientific-Technical Analyses of Impacts, Adaptations and

Mitigation of Climate Change. "Its Technical Support Unit (TSU) head was Richard Moss, who worked with me later to create consistent language in scientific assessments to address the inevitable uncertainties in evaluating complex topics. Working Group III was headed by James P. Bruce and Hoesung Lee, and its section of the report was called "Economic and Social Dimensions of Climate Change." Chapters within each section's report were written by specific authors and co-authors and reviewed rigorously by experts and governments in two draft stages before the final review. Incidentally, Jim was review editor for my chaper in the Third Assessment Report six years later.

I experienced firsthand the fraught attempts to resolve disagreements in the earliest stages of the assessment reports so we could reach consensus during the limited time of the plenary meeting. In 1994 I went to Sigtuna for the first meeting in Sweden. Previously, I had been only a long-distance author. This was my first experience with a working group lead authors' meeting, and it was more formal and hidebound than I expected. The way some IPCC detractors tell it, the IPCC is a loose cannon of radicals, which is nonsense—it is much more a debating club dominated by conservative empirical scientists. It took several assessments just to get them to mention surprises, let alone formal subjective probabilistic treatment of uncertainties.

I had published a chapter on surprises and nonlinearity—responses of a complex system that are not necessarily in proportion to the amount of a stimulus that causes the response—in a book in 1994 that John Houghton didn't like—he thought nonlinear surprises were too speculative and would be abused by media—a debate we had at a meeting in Oxford several years earlier. "Aren't you worried that some will take this surprises issue too far?" he asked. "I am, John," I replied, "but I am equally worried that if we don't tell the political world the full range of what might happen that could materially affect them, we have not done our jobs fully. If we choose to withhold discussions of

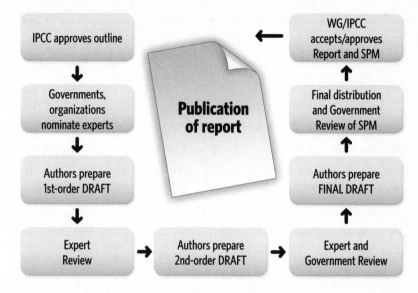

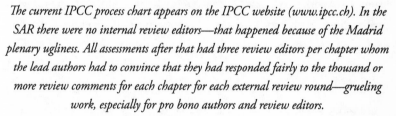

The current IPCC process chart appears on the IPCC website (www.ipcc.ch). In the SAR there were no internal review editors—that happened because of the Madrid plenary ugliness. All assessments after that had three review editors per chapter whom the lead authors had to convince that they had responded fairly to the thousand or more review comments for each chapter for each external review round—grueling work, especially for pro bono authors and review editors.

major controversial issues like surprises until we have greater consensus, then we are substituting our values on how to take risks for those of the society—the right level to decide such questions." John concurred, but remained concerned about abuses.

I had long argued that we cannot be precise on future inferences and needed a formal treatment of uncertainties. In Sigtuna I spoke up at the meeting for Working Group I on advancing our understanding of the science and mentioned two studies, one published in 1994 by William Nordhaus and another one by M. Granger Morgan and his postdoc David Keith that was scheduled to be published in a year or so. Both had asked a group of experts to provide statistical subjective probability distributions for how much damage, in the form of lost GDP, climate changes of various degrees of warming would cause (Nordhaus) and how much warming a doubling of CO_2 would create (climate sensitivity, in the Morgan and Keith paper).

I argued that we shouldn't have statistical distributions just for the damage functions. We should have them for climate sensitivity as well. We should have statistical distributions based on the uncertain parameters on the carbon cycle model. We should also have statistical distributions for how humans will behave over the next hundred years if they continue to emit carbon dioxide. We should end up with joint probability distributions and put them all together and, I concluded, that's the final "answer" that we should be giving to the political world.

That idea was viewed as radical by most of my climate science colleagues in Working Group I. They asked, "How could we come up with probability distributions in the future when we have enough trouble understanding the present?" The argument that I kept using over and over again was: "If we don't assign probabilities to climate sensitivity and to other things that we have expertise in, then the economists who will build models based upon the ranges that we give will have to *guess* what they think we thought the probability distributions are." Yet I had no allies willing to come forward at that time.

In the end, though, despite the worry of John Houghton that discussions of surprises could be taken out of context by extreme elements in the press and NGOs, we were able to include a small section on it. In the forward-looking chapter 11, since it addressed what we should do later, we discussed the need for more formal and subjective treatments of uncertainties and outright surprises. The more rapidly a nonlinear system is forced to change by greenhouse gases or other stresses, the more likelihood to have surprises. That is essentially what I said that led to the very last sentence of the IPCC Working Group I 1995 Summary for Policymakers (SPM). It was fashioned in Madrid at the plenary held one month before the final plenary session to accept all three working group reports. I wrote the sentence on surprises at two o'clock in the morning—after a delegate from France (glaciologist and climate modeler Jean Jouzel) said, "There's nothing on surprises in this report."

I told him what he already knew. "That's not true. The chapter on advancing our understanding, Chapter 11, had a paragraph that made this point."

He said, "Well, if it's in there, France thinks it belongs in the Summary for Policymakers." It was a good thing that France had the foresight to send a top-notch scientist who knew the subject—rather than only diplomats—to represent them at the plenary.

As contentious as that meeting was, we were somehow able to avoid argument on that issue. "Nonlinear systems when rapidly forced are particularly subject to unexpected behavior" was the Summary language that was approved by the delegates. What surprised me was that the Saudis, after getting an approving nod from Don Pearlman, the guru of contrarian organizations, agreed. They wanted to frame climate change as largely uncertain, and what is more uncertain than stressing surprises? Privately, Bob Watson asked me the same thing. "Won't that let them say we don't know anything?"

"But Bob, if the phrase 'when rapidly forced' stays in, then the obvious policy conclusion later on is 'don't force the system so rapidly that we risk nonlinear surprises.'" Bob understood instantly and the United States joined with the Saudis and the French and all the others—we got the line through in the wee hours, since it had backup research in Chapter 11.

The battle in Madrid was really over Chapter 8—headed by Tom Wigley, then at NCAR, and Ben Santer at the Lawrence Livermore National Laboratory. Ben had led a team of scientists for several years to look for "fingerprints" of the causes of climate change in the observational record. The patterns were not supposed to be just observational, but predicted by climate models driven by different causal forces. These causes included natural factors such as sunspot activity and volcanic eruptions, and human forcing such as aerosols, ozone depletion, and greenhouse gas increases. The models predicted that ozone depletion would cool the stratosphere, mostly near the high latitudes. However, greenhouse gases would not only cool the stratosphere everywhere, especially in the tropics, but also warm the lower atmosphere and the surface. If "the sun did it," as was so often asserted by skeptics, then a change in the energy output of the sun would warm the stratosphere and troposphere simultaneously.

Santer, Wigley, and colleagues found that anthropogenic climate changes could be detected from a number of independent "fingerprints," especially the observed cooling of the stratosphere and the warming of the lower atmosphere. The mere detection of a warming trend in the lower atmosphere was not enough for a smoking gun, but we also needed to know the source. Santer and company said human forcings were at least partly culpable, and this was a smoking gun of "discernible" influence. The finding set off a firestorm.

The Saudis objected loudest, saying it was completely unacceptable to their delegation—meaning they would not allow consensus. Joined by Kuwait, they entered objection after objection. Lenny Bernstein, from oil interests at that time, kept running messages from Don Pearlman to Mohammed Al-Saban, the Saudi chief negotiator, which led to more and more technical challenges. At one point we in the lead authors' sections were taking bets on how many minutes would pass after Bernstein handed a note to the Saudi chief negotiator before the Saudis raised their flag with yet another objection. Santer and Wigley were on the grilling podium defending their detection and attribution chapter for a long time. Meanwhile, the Saudis appeared to be encouraging a Third World rebellion against the entire Chapter 8, and at one point a lone delegate from Kenya rose to endorse dropping the entire chapter. Bert Bolin then stepped in and called for a "contact group" to meet immediately to work out the problems and report back with acceptable language. In the past, such smaller groups would eventually work out language all could live with, and it would be accepted by the plenary and the text gaveled into the record by the chair.

Contact groups were an important part of an IPCC plenary. In a room with 120 nations and 500 people, delegates had sufficient anonymity to say any damn fool thing they wanted to blather, regardless of how unscientific it might be. But in a contact group—eyeball to eyeball across a small table—you couldn't simply assert national positions as truth when you know you will be shot down by others who know the science and are sitting just a few feet away. So the small groups usually work. A dozen of us, headed by Martin Manning of New Zealand, appointed by the chair, went off for the day to address the Chapter 8 language.

Contrarians like Fred Singer sat in the back of the room, but surprisingly, Fred didn't say too much. The scientists there to defend Chapter 8 were of course Ben and Tom, joined by Mike MacCracken, who was a U.S. delegate, and me. Several delegates from developing countries who were following the Saudi line came, but no Saudis showed up, to my surprise. The Kenyan who had called for the chapter to be dropped by chance sat down right next to me. He scowled a bit at first, but it became clear that he was a trained meteorologist, but not an expert in climate modeling, so he needed a scorecard. So I started to explain to him how parameterizations worked, what detection means—just noting a statistically significant trend—and what attribution meant—attributing it via fingerprint methods to certain causes. He was smart and open, and after a few hours he got the gist of the arguments, asking a few relevant and useful questions.

When he first arrived, clearly not knowing the detailed science, I wondered how we could fulfill Bert Bolin's vision of entraining those from developing countries into the process. But it soon became clear to me that this delegate was not there as a front-rank scientist, but as an adequately trained witness to the proceedings. His role would be to certify the openness and fairness of the process, not to pass expert judgment on all the details of the complex scientific arguments. I couldn't tell what he really thought, although after many hours his scowls were gone and he was a productive member of the debate.

Well past dinner, we finally hammered out a rough agreement that the chapter was right, but we should frame any positive conclusions about attribution in a swath of caveats—as was already done in the chapter. In fact, it was overdone, given the usual propensity of scientists to lead with their caveats and not their message.

So back to the plenary we went. The compromise text was explained by Martin Manning, and the chair called for agreement. The Saudis objected, giving altogether new objections, which they probably got from Pearlman. Ben Santer is mild-mannered, careful to a fault, and polite. But he lost it. "You keep objecting," he told chief delegate Al-Saban, "but you didn't even send anybody from your delegation to witness the proceedings. They could have objected in the contact group and gotten answered by the scientists there."

Al-Saban slammed his fist on the desk, saying, "I am a representative of a sovereign nation and you are just a scientist. Our delegation has many issues to address, and we didn't think this was as important as other things. And besides, we're just a small delegation and can't attend everything." Then the delegate from Kenya raised his flag. My stomach was in knots, for if he continued to back the Saudis, it would have precipitated enough developing countries to kill the most important new finding of the IPCC.

The delegate spoke modestly. "I too am from a small delegation, but unlike the Saudi Arabian delegation, I am the only member of my country here. But I was convinced by the Saudi concerns that this was the most important issue we faced, so I went to the contact group for hours, and when I understood what was happening, I became fully convinced the lead authors had correctly portrayed the science, and thus I withdraw my objection to Chapter 8 and move we accept the text from the contact group."

You could have heard a pin drop. The Third World revolt was killed gently and decisively. Although Kuwait and the Saudis still refused to allow consensus, I told John Houghton privately later, "Let them not sign. They will become a world laughingstock, these OPEC special interests, claiming to know more science than the lead authors and the rest of the world. Just give them enough rope—they'll hang themselves." John agreed, but it would be Bert's decision.

Later on Bert negotiated the "discernible" word heard round the world, and the Saudis and Kuwaitis left the meeting with an asterisk, saying they did not accept that paragraph on discernible impacts of human activities on climate. But as predicted, in the full plenary a month later to accept all the working group reports, mysteriously— and without fanfare—the Kuwait and Saudi Arabia objections disappeared, so the final report had consensus after all. It was a stunning success and produced the single most important paragraph in assessment history—it probably led to the viability of the Kyoto Protocol process, in my view. The paragraph in full reads:

Our ability to quantify the human influence on global climate is currently limited because the expected signal is still emerging from the noise of natural variability, and because there are uncertainties in key factors. These include the magnitude and patterns of long-term natural variability and the time evolving pattern of forcing by, and response to, changes in concentrations of greenhouse gases and aerosols, and land surface changes. Nevertheless, the balance of evidence suggests that there is a discernible human influence on global climate.[2]

Two brief final notes. About five lines of this paragraph heard around the world are caveats, and only less than two lines comprise the famous "discernible statement." An important transition word, "nevertheless," conveyed the idea that despite the required caveats, there really was a new and important message here. I am happy that I got my two cents into this paragraph by suggesting the addition of "nevertheless."

Later on, at a reception, I bumped into my new friend from Kenya and told him what he did was a great act of courage and scientific logic. Form a hypothesis, test it by going to the contact group, reformulate it based on new data, and admit why you were wrong and how the new evidence changed your position. He smiled and thanked me for that. "You know," he confessed, "when I came here, my government said, 'Don't let the meeting hurt our interests.' But I knew they really didn't care much about this process, so I just said to myself, I'll learn what is going on and make up my own mind. In truth, at home they won't even notice." It was a great act of integrity and helped change the world.

Bert Bolin was indeed right about needing all the countries in the game, not just the elite scientists. To gain international credibility, the process must involve witnesses that many nations and groups can trust. The combination of expertise and witnessing the legitimacy of the process is what has made IPCC so effective.

Those skirmishes during the IPCC meeting sessions were immediately followed by an all-out public attack on Ben Santer when the Second Assessment Report was released in June 1996. Ben was accused of personally altering the accepted language of Chapter 8 and deleting more than 15 sections of it in a blatant attempt to "remove hints of the skepticism with which many scientists regard claims that human activities are having a major impact on climate in general and on global warming in particular."[3] Frederick Seitz, his accuser, insisted that the published version of the Second Assessment Report (which would influence critical decisions on energy policy and "have an enormous impact on U.S. oil and gas prices") had deceived the public and the policymakers and that Ben Santer had violated the procedures that kept the IPCC "honest."[4] Seitz's opinion piece in the *Wall Street Journal* on June 12, 1996, was prompted by the advocacy group founded in 1990 by another climate skeptic, Fred Singer, called the Science and Environmental Policy Project (SEPP), of which Seitz was the chair of the board of directors.

The IPCC lead authors were outraged. Seitz, a former president of the National Academy of Sciences and founder and chair of the conservative think tank George C. Marshall Institute (an NGO advocate for Edward Teller's Strategic Defense Initiative—known popularly as the Star Wars program), had accused the IPCC of tampering with a peer-reviewed scientific report for political purposes. It was no surprise that he focused on Chapter 8, which contained this key conclusion. Seitz was no stranger himself to the practice of conspiracy and suppression; he had long-established links to the tobacco industry. His attack on Ben Santer was supported by similar claims made by the Global Climate Coalition (GCC), the coalition of fossil fuel industry supporters.

Ben immediately responded with a letter written to the *Wall Street Journal* signed by 40 prominent scientists who had contributed to the IPCC report. It was a clear explanation of the IPCC rules of procedure and a strong defense against the accusations of impropriety. The letter explained that the changes made after the Madrid meeting in November 1995 were in response to written review comments received in October and November from governments, individual scientists, and nongovernmental organizations and during the plenary sessions at the Madrid meetings.

"Dr. Seitz is not a climate scientist," Ben wrote. "He was not involved in the process of putting together the 1995 IPCC report on the science of climate change. He did not attend the Madrid IPCC meeting on which he reports. He was not privy to the hundreds of review comments received by Chapter 8 lead authors. Most seriously, before writing his editorial, he did not contact any of the lead authors of Chapter

8 in order to obtain information as to how or why changes were made to Chapter 8 after Madrid." Neither did Seitz contact the chair of the IPCC, Bert Bolin, or the two co-chairs of Working Group I. However, the *Wall Street Journal* editor omitted that sentence in the letter as published on June 25. More tellingly, the editor also deleted the next sentences: "Scientists examine all items of evidence before drawing conclusions. They generally avoid making pronouncements outside their own areas of expertise. Seitz has failed on both counts, and his conclusions are incorrect."[5]

That editor also deleted all 40 of the distinguished scientists' names and affiliations who had co-signed the letter, including mine.

That same edition carried a letter from IPCC chair Bert Bolin, Sir John Houghton, and Luiz Meira Filho (the co-chairs of Working Group I) defending Santer as thorough and honest and in full scientific control of the draft of Chapter 8, with which they were completely satisfied. From this letter the editor deleted five out of its seven paragraphs before publication.[6]

Seitz's "A Major Deception on Global Warming" was followed in the same newspaper by an article by Fred Singer called "Coverup in the Greenhouse?" on July 11. Singer claimed that an editorial published in the international scientific journal *Nature* on June 13 confirmed his claims of politically driven distortion and impugned the motives of SEPP.[7] The IPCC's reputation was being dragged through the mud for the specific purpose of distracting the public and the policymakers from its key finding on the link between human activities and climate change.[8]

The UCAR quarterly newsletter for Summer 1996 carried a special insert: an open letter to Ben Santer from the executive committee of the American Meteorological Society and the trustees of the University Corporation for Atmospheric Research (UCAR). Expressing gratitude to Santer and the other IPCC scientists for their work, the writers comment that Seitz's

Wall Street Journal essay is especially disturbing because it steps over the boundary from disagreeing with the science to attacking the honesty and integrity of a particular scientist, namely yourself. There appears to be a concerted and systematic effort by some individuals to undermine and discredit the scientific process that has led many scientists working on understanding climate to conclude that there is a very real possibility that humans are modifying Earth's climate on a global scale. Rather than carrying out a legitimate scientific debate through peer-reviewed literature, they are waging in the public media a vocal campaign against scientific results with which they disagree.[9]

The duplicity and unfairness of this attack on Ben Santer only served to confirm that the fossil fuel industry would use every avenue to discredit the evidence of the "discernible influence" of humans on global warming. If you can't beat the science, challenge the process—the strategy was as simple as that. Ironically, because of his long record of creative research, Santer received a MacArthur Fellowship, the "Genius" award, for his efforts a few years later. As my friend Paul Ehrlich likes to say—and I recounted this to a smiling Ben—"Doing well is the best revenge." Yet his ordeals were not over. When you hurt powerful interests, to protect their market share or ideological preconceptions, anything goes.

Around the time of the Kyoto Protocol debate, a year or so after the Santer affair, Seitz struck again. This time he used an old saw of contrarianism—get "scientists" to support your skeptical position in greater numbers than the supporters of the mainstream views. In this "all scientists are created equal" framing, he led a massive polling exercise, in which a never published paper by contrarians was sent to hundreds of thousands of technically trained people, mostly members of scientific societies. The overwhelming majority of them had never published a paper in climate science. Along with Seitz's cover letter and the contrived skeptical paper, they were sent a petition asking if they supported or rejected the basic climate change conclusions and the Kyoto Protocol. The sponsoring group, the Oregon Institute of Science and Medicine, harvested about 17,000 signatures skeptical of the science and the Protocol out of an unknown number of solicitations—many hundreds of thousands, for sure. They then claimed that many more "scientists" supported skeptical interpretations than mainstream views—and the media dutifully carried the "news," and favorable members of Congress advertised it. When a science writer called for my reaction, I asked him how many names he recognized as legitimate climate scientists. He said a handful—the usual suspects. My response was tongue in cheek. "So if all scientists are created equal, then all MDs are likewise equivalent. So I'll ask my podiatrist to prescribe my heart medicine and ask my cardiologist—who hasn't touched a scalpel in 30 years—to take off my bad toe nail. My point, of course, is these are not climate experts and are not qualified to have meaningful judgments on climate science, as they do not represent a community expert in the details of climatology. A petroleum geologist can no more tell us about cloud feedback than a climatologist could competently tell us about oil reserves. They should be ignored—don't even cover them, is my opinion." The reporter didn't take the last bit of advice but did use some of the facetious quote.[10]

Ben Santer, destined to be crucified more than once, became the victim of legalistic harassment in 2008 over a paper that he had written with more than a dozen colleagues on climate models in the tropical troposphere, published in the *International Journal of Climatology* in 2008. The deniers didn't like it because it destroyed their case. They attacked Ben again with all their strength, knowing that he would need to use his time and resources to fight their accusations, rather than pursuing his research and solidifying his findings. It was an intimidation trick in the name of due diligence.

We've seen a number of these "FOIA attacks," where individuals or organizations employ the Freedom of Information Act to request data such as line-by-line programming codes for climate models or data analyses from certain scientists whose work the opponents wish to malign or impede. Ben was a targeted victim, similar to Michael Mann, whose "hockey stick" graph of the reconstruction of temperatures over the past millennium showed in 1999 that the 1990s had been the warmest decade in a thousand years.

Mann was attacked by climate skeptics using a similar legalistic tactic, demanded by a denier congressman from Texas, Joe Barton, chair of the Committee on Energy and Commerce. It took a full review of the hockey stick study by a National Academy of Sciences committee to prove that although Mann and colleagues did make some minor errors—as is normal in creative, original science—the basic conclusion, that the past several decades were the warmest in at least 500 years, remained. In 2007 the IPCC Fourth Assessment Report SPM said about this controversy: "Average northern hemispheric temperatures during the second half of the 20th century were very likely higher than during any other 50-year period in the last 500 years and likely the highest in the past 1300 years."

Ben Santer, still at Lawrence Livermore, did not want to spend his time making his line-by-line computer program accessible to public perusal. He knew—and obviously so did his attackers—that the programming codes would be virtually useless to anyone trying to replicate his results. The beauty of systems science is that we come to conclusions through independent efforts that confirm one another—it's not merely a matter of rerunning someone else's computer code or models. We like independent groups using independent models coded by each separate group to try the same experiments or look at the same data set, and if reasonably conforming, we increase our confidence in the conclusions. Personal codes are so idiosyncratic to the programmers that it could take months to explain them to others who could, in much shorter time, do an independent audit by building their own code using the same equations or data sets. The National Science Foundation has asserted that scientists are not required to present their personal computer codes to peer reviewers and critics, recognizing how much that would inhibit scientific practice.

A serial abuser of legalistic attacks was Stephen McIntyre, a statistician who had worked in Canada for a mining company. I had had a similar experience with McIntyre when he demanded that Michael Mann and colleagues publish all their computer codes for peer-reviewed papers previously published in *Climatic Change*. The journal's editorial board supported the view that the replication efforts do not extend to personal computer codes, with all their undocumented subroutines. It's an intellectual property issue as well as a major drain on scientists' productivity, an opinion with which the National Science Foundation concurred, as mentioned.

Ben Santer was on the verge of quitting the Livermore lab over this pernicious attack—it would have been a great loss to climate science. He worked for the Department of Energy, and it would have been admirable if the DOE had stepped in and informed McIntyre that his request was unreasonable. DOE, still under the Bush Administration at that time, did not. The Obama Administration, however, appears poised to fight these battles side by side with concerned scientists. Time will tell how it turns out.

SETTING FIRM GUIDELINES FOR UNCERTAINTY

The following year after SAR was published, Richard Moss and I decided to address the question of how uncertainties were treated in IPCC's Second Assessment Report and come up with more concrete guidelines. Both of us had independently tried in our respective working groups to apply more formalistic treatments of uncertainties. How certain were we about the "discernible influence"? Was one group's "very certain" actually the same level of confidence expressed by another group's "extremely possible"? How about high confidence versus medium confidence? There was no consistent guidance.

I said, "Richard, we need to join forces. There will be a third assessment coming up in 1998, and we have to confront our colleagues with the fact that they cannot continue to duck the question of subjective statistics. Because if they don't, then either the economists will do it for us in their models or, worse, the politicians."

Political leaders *want* us to tell them what can happen, and what are the odds—e.g., Bill Bradley. If we don't tell them, they have to guess. Richard and I decided that we should convene a group of lead authors from all three working groups, and address the question of how uncertainties were treated in IPCC's Second Assessment Report. I suggested that we go to John Katzenberger of the Aspen Global Change Institute (AGCI) and have a two-week summer meeting on exactly that topic. Katzenberger liked the idea and was able to convince the U.S. Global Change Research Program to fund it.

In the summer of 1996 we held a session on uncertainty. We also invited Granger Morgan from Carnegie-Mellon, arguably the most significant person in bringing the discipline of decision analysis to the environmental community, and Elisabeth Paté-Cornell from Stanford. They were decision analysts, formally involved in how to make decision-analytic protocols. Morgan emphasized that the worst thing anyone writing an assessment report can do is use terms such as "likely" and "confident" without linking them to a quantitative scale. "Every person has a different probability in mind when they use those words, and therefore we all will be talking past each other."

Richard and I found a hundred examples in the IPCC reports where exactly that inconsistency had occurred. Granger knew what he was talking about. He once informally polled the EPA science advisory committee, of which he was a member, on what they thought the probability of "likely" meant in a made-up example of a cancer risk. These trained experts often disagreed by nearly a factor of ten.

We evolved quantitative scales to define terms such as "likely," "unlikely," "high confidence," "medium confidence," "low confidence," and so forth. We fought it out for two fun weeks with the climatologists, engineers, ecologists, statisticians, economists, and journalists present, and sent the report to Bob Watson, who then invited us to talk to several IPCC groups. The idea was controversial—we said that this topic crosscuts all three working groups.

It turned out that Bob and others thought we should employ multiple crosscutting themes. One of the leaders of this crosscutting effort was Rajendra Pachauri of India, later to become the next general chairman of IPCC. One crosscutting theme would cover uncertainties, and another would address sustainability and equity, a concern that the Third World countries were demanding. The IPCC no longer wanted to use neoclassical economics with its cost benefit calculus in which a dollar lost in the Third World was to be balanced by a dollar gained in the rich countries and the situation called "welfare neutral." IPCC representatives, especially from developing countries, wanted distribution and equity to be an explicit variable. They also wanted costing methods to be a crosscutting theme, because after all, you don't just cost things straight into markets—you have to deal with the loss of lives and heritage sites and equity and accountability based on who caused more of the problem.

In total, four of these guidance papers were prepared, and Moss and I were asked to write up the first one on uncertainties. We worked about a year and a half on this draft, building it on the invaluable experience of thrashing out the issues at AGCI the previous summer. Interested lead authors from all three working groups followed with several rounds of reviews by e-mail. In 1998-1999, when IPCC was meeting for the Third Assessment Report, we sent out the drafts to the working groups. We negotiated a quantitative scale—we would define "low confidence" as a less than 1-in-3 chance; "medium confidence," 1-in-3 to 2-in-3; "high confidence," above two-thirds; "very high confidence," above 95 percent; and "very low confidence," below 5 percent, for example.

It took a long time to negotiate those numbers and those words. Some people still felt that they could not apply quantitative scale to issues that were too speculative for real scientists. One nasty commenter said that "assigning confidence by group discussions, even if informed by the available evidence, was like doing seat-of-the-pants statistics over a good beer." He never answered my response—"Would you and your colleagues think you'd do that subjective estimation less credibly than your Minister of the Treasury or the president of the U.S. Chamber of Commerce?" It was, sadly, the same argument as at the CIAP meeting in November 1972.

So we had two scales, a quantitative one defining the probability ranges for words such as "likely" and a qualitative four-box verbal scale. We used phrases such as "well established" if a finding was backed by solid data and agreement between theory and data. We used words such as "speculative" when there wasn't much data or agreement. We decided on "established but incomplete" and "competing explanations" for the intermediate cases.

For the next two years Richard and I became known as the "uncertainty police." I read thousands of pages of draft material in various working groups where people employed uncertainty terms not in accordance with the guidance paper terminology or scales for consistency. For instance, they would write a sentence where they would say that because of uncertainties, we can't be "definitive." I wrote back, "What is the probability of 'definitive'?"

Other typical sentences in early drafts would say, "The range of outcomes could be anywhere from one to five degrees Celsius change." And then they would put in parentheses, "medium confidence"—that's completely incorrect. It was "very high confidence," because they were talking about the fact that between one and five degrees was a very, very likely place in which the outcome would occur. But they didn't want to say "very high confidence" because nobody felt very confident about the state of the science at the level of pinning it down to, say, one degree. So I would help people to rewrite, and say that we have "low confidence" in specific forecasts to a precision of a half degree, but we have "high confidence" that the range is very likely to be one to five degrees.

Simple clarifications like that were needed to achieve consistency, yet it took a long time to get some in the community accustomed to consistent use of phrases linked to the scales. Every time phrases like "evidence is insufficient to build consensus" occurred, I would write in big red letters: "What is the probability of a consensus? Strike this please and assess the group's judgment on the confidence in each important conclusion or process in agreed uncertainties guidance language."

Consensus is not necessarily built over the conclusions themselves, but in the confidence we have in a host of possible conclusions. With that kind of information, we can discuss risk management in an informed manner by weighing both the possible outcomes and the assessed levels of confidence for each possible conclusion. We either know the outcomes "well," "sort of," or "hardly at all." Just say what we do and don't know, and do not leave the conclusion out because it isn't well established yet. (That is policy prescriptive—an IPCC no-no.)

This approach helps policymakers, since it is the job of society, through its officials, to make the risk management decisions informed by our conclusions and accompanying confidence estimates. We had many contentious go-rounds on that issue.

Linda Mearns at NCAR was one of the few lead authors who worked with both Working Groups I and II. She served as a valuable ambassador to get some reconciliation across the different perspectives. For example, there were some physical scientists in Working Group I who were leery of subjectivity and risk management, and there were some social scientists in Working Group II who felt that society, not scientists, should choose how to take risks after all the possible conclusions were reported, not just the consensus ones. Those were not scientific differences, but value judgments of those with very different training and experiences. It took us quite a long time to get both sides to first understand and eventually respect the other point of view. My role was not to endorse one or the other, but rather to be sure all our reporting was explicit about our underlying assumptions; this gave us a "traceable account" of all the underlying processes and assumptions behind important conclusions. To this day, that process is not yet fully converged across the different disciplines of IPCC or the scientific community in general.

In the end, the guidance paper was very popular with the governments that attended the plenary sessions—perhaps they welcomed it more than the scientists who provided the assessments. Each Summary for Policymakers contained an explicit description of the quantitative links and terms such as "likely" or "high confidence." The Working Group I and II reports in the Third Assessment Report addressed these issues for the first time in a fairly consistent and quantitative way. Unfortunately, Working Group III sat on the sidelines of this debate, using whatever language the lead authors, the reviewers, and the plenary approvers could live with. I was disappointed that they wasted the opportunity for better assessment conclusions for policymakers, but that problem was corrected to some extent in the next round. It takes time to change culture—even in very "sophisticated" scientific communities—as Margaret Mead explained to me decades ago.

Our job was not done. The Third and Fourth Assessment Reports were to assess the subjective probabilities for events a hundred or more years in the future. The analysts' reluctance to do that is understandable. Imagine someone in the Victorian era being asked to assign probabilities to what the emissions are right now. You might find the exercise laughable. On the other hand, how is a political leader to know how much of a resource to invest in mitigating a problem if you don't have some idea about the likelihood of its escalation?

So I continue to call for a double strategy where appropriate, whereby you provide not only subjective probabilities, but also the degree of confidence that you feel in those forecasts or in the underlying scientific processes involved. Our degree of confidence will be low on many of these forecasts, unless the ranges are broad enough. I think that approach is more responsible than ducking the question of confidence altogether, and leaving it to every polemicist who wants to grab the IPCC numbers out of context to support a radical view. That's already happening, as people are grabbing these "equally sound" story lines from the Special Report on Emission Scenarios published in 2000, because IPCC never constrained them by suggesting what the experts believe to be the relative likelihood of each of those either modeled sensitivities or emission scenarios. We have plenty of work yet to do to convey in consistent language the degrees of our knowledge and reasons for concern.

THE CONTENTIOUS KYOTO PROTOCOL

In December 2009 the world is meeting in Copenhagen to work out plans for addressing the issues of climate change after the Kyoto Protocol expires in 2012. It's a welcome step forward in addressing the risks of anthropogenic global warming and the mitigation and adaptation strategies that are required to sustain life on our planet.

The Kyoto Protocol emerged from the 1992 Rio Earth Summit, where most of the nations of the world agreed to target reductions in their emission of six greenhouse gases. Technically, the Kyoto Protocol is an amendment to the UN Framework Convention on Climate Change (UNFCCC). After five years of international wrangling prior to the December 1997 international summit in Kyoto, Japan, world leaders for the first time agreed to specific targets and rough timetables for reducing greenhouse gases. Amid all the contention and frenzy accompanying this event—attended by thousands of journalists, lob-byists, diplomats, and observers (as well as a few heads of state)—you needed a scorecard to keep up with all the points and counterpoints surrounding the negotiations.

In the pre-Kyoto cacophony were at least five "sides" vying for political momentum. The elements of many of their positions con-tained important components of an eventual global climate treaty, but each aspect neglected many other components.

The fossil fuel industry was a big voice. Their active industry media campaign, most visible in the United States, was designed to frighten pol-iticians and the public into believing that any global warming treaty to emerge from Kyoto that did not include the developing countries would not slow global warming and would bankrupt participating nations. The campaigns were facilitated by industry-sponsored NGOs, some of which were present at Kyoto. Moreover, their carefully crafted, highly profes-sional television and newspaper advertisements asserted that the Kyoto Protocol would unfairly disadvantage U.S. products in international competitive markets by raising the energy costs embedded into Amer-ican goods but not those produced in most Asian and other Protocol-exempt competitors. This lack of globally uniform participation would favor jobs abroad at the expense of American workers, they said.

In the summer of 1997, the U.S. Senate, spurred on by this indus-try effort, overwhelmingly passed a resolution calling for the United States not to enter into any climate treaty that exempted develop-ing countries from mandatory participation. Because two-thirds of the Senate must agree to ratify any treaty signed by U.S. diplomats before it can become legally binding on the United States, this was a major stumbling block that was never resolved. The United States and Aus-tralia did not initially ratify the Kyoto Protocol—until, in the case of Australia, the Labor Party displaced the conservative government and signed the Kyoto Protocol as one of its first official acts in 2008. The

Kyoto rules did not even go into effect until February 2005, after Russia finally ratified in late 2004, providing the required number of ratifications to account for an average of 55 percent of global emissions. That would have, if all signatories followed the agreement, cut emissions in 2012 to about 5 percent below the 1990 level.

The yin for the industry's yang were the environmentalist NGOs, whose sights were set on large reduction targets for CO_2 emissions by 2010. They were demanding strong targets and near-term timetables because, unless emissions dropped soon, the world would see a doubling of CO_2 by 2050. The environmentalists were angry with the coal and oil industry for sponsoring what they saw as a deceptive and self-serving media scare campaign, and they were annoyed with the U.S. government for proposing the weak target of returning emissions only to 1990 levels by 2008-2012, whereas the European Union (EU) was proposing much bigger cuts. In practice, the reductions didn't happen. Today the levels of CO_2 have increased by roughly 15 percent and many climate changes are happening even faster than predicted. It didn't happen since this Protocol was, as I said over and over again when I was at Kyoto, "targets without teeth." There was no consequence to a country for not keeping its commitment. It was like having traffic lights and speed limits without cops and judges—voluntary compliance will certainly not be anything like it would be if the rules were enforced. Without concerted action, CO_2 equivalent levels since 2000 are currently on track to triple preindustrial levels by 2100.

The Europeans proposed a significant cut for the EU as a whole and had sharp rhetoric for the United States and Japan—large emitters who appeared unwilling to join the EU in sacrificing for the sake of the planet. They had originally agreed that the less developed countries should be exempt from this initial round of reductions, as they needed time to develop and catch up with the more developed countries, which historically had created the bulk of the problem. The

United States had agreed to this earlier but reversed its decision in the months before Kyoto.

The developing world saw itself as victim once again. Do nothing, and they get hurt from global warming, because of their lack of resources for adaptation and relatively warm starting conditions. Cut emissions globally, and they get hurt most because it limits their plans for economic development through cheap carbon-based energy sources. Many representatives regarded advocates of climate change policy as part of a colonialist-like conspiracy to keep industrialization down in the developing world so that richer countries could maintain their favorable economic positions. They demanded exemption from emission cuts until their economies approached the strength of the highly developed nations in per capita emissions terms.

The U.S. position going into Kyoto acknowledged the fossil fuel industry's argument that the global warming problem required a global solution and called for participation of all countries. They called for an economically efficient way to achieve lowest cost emission reductions through a tradable permit system for CO_2 emissions and joint implementation. They also called for subsidies to develop alternative, less carbon-intensive energy technologies. The U.S. position to allow some parties to pay for cuts outside of their borders was criticized as an attempt to escape domestic emissions cuts. The United States responded that as long as they met their quotas, why shouldn't they do it where it is cheapest—at home or abroad, depending on circumstances.

The conference took place December 1-11, 1996. I was there for every minute of it—although I did go shrine hopping and bird-watching one day in the beautiful city of Kyoto. Formal sessions in which statements from national delegates and a few heads of state—the prime minister of Japan and Vice President Al Gore, for instance— took place in a gigantic hall. While packed for the opening speeches by high government officials, most of the time it was sparsely populated, because each delegate's speech could be observed in several languages at one of the many television monitors within the massive conference center. While speeches were being delivered, interviews, press conferences, private negotiations, strategy sessions, and political negotiations were going on simultaneously.

One particularly visible environmental NGO was the Global Commons Institute from the United Kingdom. Its charismatic leader, Aubrey Meyer, became a darling of developing countries by pushing for contraction and convergence. This called for the overall planetary emissions to contract to much lower levels by mid-century and for the low per capita emissions in poor countries to converge with the higher emissions in rich countries, as a measure of equity. This analysis was not popular with economists, since it was basically an idea presented via great graphics, but it was not based on an accepted economic model calculation making costs and benefits explicit. Regardless of its merits, it is a good example of the kinds of ideas that were kicked around in the informal sessions held before the governments got together in closed-door sessions to hammer out protocol language for the Conference of the Paarties (COP). Many of these events were well covered by the international media.

Even before Vice President Gore had finished his speech explaining the U.S. position to delegates, some environmentalists were organizing press conferences to denounce the U.S. reduction targets as woefully inadequate. Gore departed from his prepared text to instruct the U.S. delegation to be "more flexible" in their negotiating positions. Apparently, this flexibility contributed to an agreement after several round-the-clock negotiating sessions in the last two days of the ten-day event, and a compromise emerged.[11]

The Kyoto Protocol was refined over the next few years, but the United States repulsed any pressure to bring the agreement before the Senate for ratification. Gore was not able to force the Clinton Administration to go any further, particularly given the hostility in the Senate to a treaty that did not obligate the Chinese or Indians to take emission-cutting targets too. When George W. Bush took over the White House in 2001, he firmly renounced the Kyoto Protocol, despite election campaign promises to regulate CO_2 as a pollutant. In March 2006, just a year after the Protocol finally became international law for those who ratified it, he commented, "I told the world that I thought that Kyoto was a lousy deal for America." Needless to say, he never mentioned that the United States had nearly the highest emissions per capita in the world and was historically the largest accumulated polluter of greenhouse gases in the world.

WALKING THE TALK

At a time when the U.S. auto industry is down—possibly for the count—in part because of their stubborn refusal to build cars that are energy efficient, it might seem less than kind to compare that industry to one of its Japanese counterparts, Honda.

In 1997 I had a pivotal meeting with Honda executives in San Francisco. I wanted Honda to invest in producing a hybrid car. David Raney, an executive at American Honda in the Los Angeles area, whom I had worked with on an education project on transportation and knew was a straight shooter, invited me to meet with two high officials of Honda and "brief them on the latest climate science." But we understood I also was going to ask them to stop wavering and start competing with Toyota in the hybrid car market. Years earlier, Amory Lovins, chief scientist of the Rocky Mountain Institute and author of many books on energy efficiency and renewable energy, had convinced me that the technology could be profitable and environmentally friendly, and I tried to convince the executives that it was a blockbuster opportunity. Honda could help do their bit to send the right message, I told them.

The Honda officials present weren't particularly interested in saving the world. Honda was interested in selling cars in America. When

I jokingly offered to buy the first hybrid Honda shipped to America, the executives weren't terrifically impressed by my market pulling power. They were more attracted to the idea that Honda could preserve technological leadership in Japan.

"Why should we invest hundreds of millions in this," the exec asked, while the senior engineer from Honda looked on skeptically, "when we have no assurance it will return on our investment?"

"Does being one of the automotive technology leaders matter to you," I rejoined, "or will you just concede it to Toyota?"

To my surprise the engineer, by now somewhat tipsy from good California Cabernet, smiled and cheered, "Yeah, yeah Honda!" like at a football game. He quickly sobered up and pointed out that the learning and prestige of being a leader would help other product lines. The executive was still cautious, but he did admit that the possibility of sprinting ahead of Toyota through innovative research and development was very enticing, and staying even with them was important for branding.

"I'll bet you'll have long waiting lists for these for a few years, at least in California," I speculated hopefully. They said they would consider it.

In the fall of 2003, I got a phone call from my Honda friend in Los Angeles. "Where is your local Honda dealer—your Honda Civic Hybrid is on the boat!" Not only was I able to demonstrate my personal support for hybrid energy-saving vehicles, I still drive that car today after six years and 52,000 miles, and it performs as well as ever. When it's time for a new car, I am hopeful that plug-in hybrids will be widely available.

This time I was happy to defer to the likes of Larry Page, Sergey Brin, and Larry Brilliant at Google to carry the message to the powers that be. My wife, Terry, and I enjoyed standing with them at Googleplex in 2008, when they announced a multimillion dollar program to promote the integration of plug-in hybrids with the Pacific Gas and Electric Company. Progress is possible.

Even traditional corporate thinking can progress. One time in the late 1970s I was invited to address my first corporate board on climate change—an agribusiness giant. They were particularly interested in my projections that future droughts were inevitable. I told them about our data, the risks, and how they might hedge. And I also told them that over the long term—half a century—agriculture would have to adapt or simply try to cope with the coming climate change that could alter the game plan. Then I turned the tables and asked them a few questions. "So what is the purpose of business?" I asked. "Growth" was the nearly universal answer.

"I never went to business school, just engineering, but I always had a crazy notion that when big businesses were first getting rolling as corporate entities, the object was to provide jobs and products to improve our quality of life." No objections from them on that. "So growth was the means to the end of improving our quality of life though growing the economy and the economic status of the people." They were fine with that too. "But now, some kinds of growth cause pollution and land degradation that is actually threatening the very quality of life business was first built to improve." No conflicts yet, just more skeptical but still engaged looks. "And if we push too far so as to even threaten the stability of our climate, then don't we have to reconsider *how* we grow so we still improve the quality of life?" Then the coup de grâce. "So when did we switch the ends with the means and wind up in this predicament: The means—growth—somehow is now defining your objective, whereas it really should be a means to the real objective, improved life. And if growth is now causing problems given the growing scale of our impacts, don't we have to change *how* we produce our wealth to keep our eyes on the original purpose: human betterment?"

They seemed stunned. One of the men said, "I admit, I never thought of it that way but you have a good point. But our company is just one small player, and there isn't much we can do by ourselves, so we need to have everybody on board so we don't lose competitive advantage."

I didn't have the presence of mind to respond then, as I have a dozen times since, by saying, "So if you favor uniformly applied regulations for everybody to maintain a level playing field, then why do you have so many lobbyists on your payroll in Washington trying to soften or eliminate these very rules?" We can't solve this set of problems without nearly everybody on board, including the corporate players. But that won't happen without enforceable and fair rules to protect the environment. Kyoto was just another example of this issue, but on a planetary scale.

People often ask me whether I walk my talk with regard to green sustainable living. So I'll answer the question here, for the record. Terry and I invested the equivalent of several years' salary remodeling our house five years back, and much of that was "green" updating: improving window efficiency and insulation, and installing an insulating white foam roof, for example. Plus, because the house can keep its moderate temperature a long time, we did not build in air conditioning. I bike the mile or so to work much of the time and take a hybrid in a carpool with Terry the rest of the time. So our domestic carbon footprint is much less than the average Californian, and with its many regulations requiring appliance and building efficiency, California is already number one in the United States in least energy use per capita. I do have a carbon problem, however—I log more than 100,000 miles of flying time in airplanes yearly, getting about the same gas mileage per seat as our hybrid gets with one person in it. So, I like to ask my students, "Is your professor a hypocrite for having over 90 percent of his carbon footprint in the skies and thus far above the U.S. average?" Immediately they defend me: "No, you might use more CO_2 than average, but your work can save many millions of tons." It would be nice to have some empirical evidence of that, but I can only hope they are right.

Similarly, I've been asked if Al Gore isn't a hypocrite because he has a big house and flies a lot. He too can use the "I cut many more tons by my advocacy than I use" line, like my students did for me.

"But if he were moral, he'd pay the same carbon tax he is advocating everybody else to pay," a skeptic in one of my audiences once said.

"He *is* buying carbon offsets at the best price available, but undoubtedly lower than the carbon price he advocates—and I do as well. It is not hypocritical. If a senator advocated we should raise the tax brackets for the rich and the bill fails, is that senator morally obligated to pay more tax than the law requires?'

"Well, no," the skeptic admitted.

"So there is no hypocrisy unless, after the bill he advocated actually got passed, he evaded it," I said. "You can't expect one individual who in good faith proposes rules for all—including himself or herself—to pay extra that nobody else has to pay. As long as they keep trying to get the rules changed, they are perfectly ethical doing what everybody else is doing in terms of paying for carbon. But they are likely doing more than most to get a real price on carbon for everybody, themselves included."

1998: A VERY BAD YEAR

The summer that followed the meetings on the Kyoto Protocol was among the hottest recorded to date. It also was one of the driest and one of the wettest, depending on where in the world you lived. Of course, one year's weather is just part of natural variability, but as warming trends build, some extremes are experienced more often. Thus, the weather that year seemed a portent of climate changes to come, even while that danger was vociferously denied by the climate skeptics.

Regions of the United States and other parts of the world experienced a variety of weather and climate extremes during June through

August. These events and conditions included drought and widespread fires in Florida, a heat wave and drought across parts of the southern U.S., flooding in China and parts of the United States, and Hurricane Bonnie striking North Carolina and Virginia. Overall costs exceeded $30 billion (including over $10 billion in the U.S.), and the death toll exceeded 3,000 (including over 200 U.S. fatalities).

In the U.S., drought and extreme heat affected an expanded area of the South, from Texas and Oklahoma eastward to the Carolinas, Georgia, and Florida. Oklahoma's Cooperative Extension Service at Oklahoma State University distributed a "Drought Survival Kit for Cattlemen." In agricultural losses (crops, cattle, etc.), Texas experienced over $2.1 billion in losses, Oklahoma about $2 billion, Florida about $175 million, and Georgia over $400 million. Overall economic costs totaled two to three times the agricultural losses. Even the bats in the Daniel Boone National Forest in Kentucky abandoned their dark-colored, unventilated bat houses in sunny locations to seek new homes in the shade; without that alternative they may have left the habitat altogether, as Mexican free-tailed bats did in Houston.

China suffered massive flooding concentrated in three areas: along the Yangtze River in south-central China, across extreme southern China in the area around the Gulf of Tonkin, and across the north near the Russian border. The heaviest reported rainfall was at Qinzhou, with an incredible 68.28 inches of rain during the June-July period. According to official Chinese government reports, 3,656 people were killed by the floods, the second worst in more than 130 years. The floods left 14 million people homeless, affected 240 million people, and caused well over $20 billion in damages.[12] Were these catastrophic events the random vagaries of nature, or the beginning of a changing pattern of extreme events linked to anthropogenic warming? Or both?

What was happening to the planet? What could we do to adapt to, if not prevent, future events like these? Would they continue to worsen as the climate warmed? At this point, we were beginning to put together the plans for the IPCC Third Assessment, using our newly developing guidelines for uncertainties. We hoped to come up with some consensus among nations about what to do about the planetary impacts.

THE IPCC THIRD ASSESSMENT REPORT

In February 2001, Terry and I flew to Geneva for the IPCC plenary session. Our task was to convince some 300 representatives of the 100 governments attending that global warming was not just a theory but was already influencing the behaviors and movements of plants and animals all around the world.

Terry had been working on this problem for over five years, and she would take her place on the dais to address some skeptical delegates who, despite being in the diplomatic corps of their countries' governments, were anything but diplomatic when it came to tackling issues affecting their national interests. And there was the $64 billion question: What was the U.S. delegation—always a positive force at previous plenary events—going to do now that George W. Bush was President? Luckily, the delegation was so low on the Bush team's list of priorities that we didn't have to face the political ideologues we feared might attend.

Having had a hand in Terry's research, I was mulling over how to deflect the more obstreperous remarks that would undoubtedly come from the representatives of the OPEC nations or big coal-producing countries. At meeting after meeting, these representatives were as disruptive as possible. They refused to accept the seriousness of climate science, and rejected solutions such as a fee for dumping greenhouse gases into the atmosphere; more efficient use of fossil fuels, thereby reducing demand; and a major increase in the use of renewable, sustainable energy. A global shift away from burning fossil fuels would cost these countries trillions of dollars, so how could anyone expect them to embrace wholeheartedly the global warming problem?

Additionally, the developing countries felt that the West had gotten wealthy by using the atmosphere as a giant free dump for their tailpipe and smokestack emissions. Some believed that the wealthier countries had invented new obstacles—global warming, the greenhouse effect, and policies to cope with them—only to block their progress. Their concerns were understandable but dangerous for the planet.

The intense three-day preplenary meeting of the lead authors took place at the modernistic World Meteorological Organization building, across the street from the Geneva office of the United Nations, the venue for the upcoming plenary. We met our colleagues—scientists, economists, governmental agency leaders from around the world—who made up Working Group II. Our responsibility was not to forecast how the climate would change (that was the job of Working Group I), but rather to ponder the questions "So what if the climate changes? How will society and nature cope?" Our goal in the preplenary session was to use governments' comments to revise the Summary for Policymakers of our draft IPCC Assessment Report. The report discussed how the world's people, rivers, crops, glaciers, plants, and animals would be affected by and forced to adapt, if they could, to climate change, and it summarized the pros and cons of suggested policies for helping them make the adjustments. As scientists, we never recommend which policies should be chosen, as that is a violation of our prime directive not to be "policy prescriptive" and is the responsibility of the decision-makers who would be attending the plenary.

The main friction during this preplenary meeting resulted from disagreements between a few climatologists, including myself, and some economists over the potential seriousness of global warming. Most climatologists in attendance believed that no one could precisely quantify the extent of the damages that climate change could impose on nature and society simply by studying markets and other items that were easily quantifiable in monetary terms (such as changed crop and forest yields and property loss due to rising sea levels). In my work with the IPCC, I wanted to value lost biodiversity and the heritage sites that would be destroyed if small island states like the Maldives and Tuvalu were flooded by rising sea levels. Personally, I feel we should take a preventive approach to climate change, reducing greenhouse gas emissions now to lower the risk of serious dangers that climate change may pose later.

These economists, however, noted that small island states represent only a tiny fraction of the world's gross domestic product, and thus their demise would have almost no impact on the global economy when considered on a "one dollar, one vote" basis. They were reluctant to assign value to anything not traded in markets and tended to disagree with using a precautionary approach (such as cutting emissions now), contending that standard economic models showed that spending significant amounts now to reduce emissions was not cost-beneficial, given that most of the benefits would only be felt far into the future, which has little present value. These economic analyses are misleading, however, since they do not tell the whole story.

The decision on whether or not to act on many issues, climate change included, often depends on what kind of error is of most concern. If governments were to follow the precautionary principle and act now, but their worries about climate change later proved to be unfounded and greenhouse gas emissions didn't greatly modify the climate, they would have committed what is known in economics as a type I error—a false positive. If governments took no hedging actions because of the uncertainty surrounding climate change issues, but then drastic climate changes *did* occur, they would have committed a type II error—a false negative. In any policy debate about climate change, I say that I'd rather risk a type I error. In other words, better safe than very sorry—but within limits, as costs do matter in a world of limited total resources. Regardless, this is a policy prescriptive issue and not a judgement IPCC should make as an assessment team.

Despite the disagreements within our group—over how, or even if, to put a dollar estimate on the value of cultures or species lost, and how to discount the loss of the Greenland ice sheet that could raise sea levels by over four meters (13 feet), but not for centuries—we tried to bury the hatchet and focus on the deeper conflicts with governments that were sure to surface at the plenary. We knew that certain representatives would wage fierce battles to try to bring the report in line with their countries' interests, even if it meant ignoring growing scientific evidence on risks of global warming.

Many of the debates would inevitably result from ambiguity regarding whether human-induced climate change was actually occurring and how severely global warming would affect the world during the next century or two. "Deep uncertainty" is the jargon I apply to situations in which both the probabilities of specified outcomes and their consequences are not well established. Climate change definitely qualifies in many areas. It wasn't until the release of the Fourth Assessment Report in 2007 that the IPCC could state with confidence that "human-induced warming of the climate system is widespread."

With such uncertainty in 2001, each special interest group could latch on to the climate outcome that best suited its position, regardless of how likely or unlikely that outcome might be. Deep ecology groups risk type I errors and fear type II, citing the most pessimistic outcomes, warning of catastrophe, and pushing for the implementation of energy taxes and other abatement policies as well as growth in renewable energy. The auto, oil, and other fossil fuel-intensive industry groups, to continue the stereotype, tend to be the extreme optimists in the global warming debate over climate effects and impacts (if not outright deniers). They ignore or downplay the potential hazards of climate change, picking the least serious potential outcomes and stating that the uncertainties are still too large to enact climate policy and that the economy could not stand the shock of a government-mandated price on carbon emissions.

When the plenary meeting at the UN headquarters began, the first and most controversial issue on the agenda was Terry's work. After analyzing the behavior of some 1,000 species of plants and animals all around the world (as reported in the 150 papers that were used to perform the analysis), she and her team were able to show that about 80 percent of the species that exhibited some kind of meaningful changes over the past 10 to 40 years or so had changed in the direction that would be expected with warming—that is, trees flowered earlier, birds migrated to their breeding grounds sooner or laid their eggs earlier, and butterflies moved up mountains or to cooler regions closer to the Poles. Thus, Terry declared, there was a "discernible impact" of recent temperature trends on plants and animals. This was a spectacular conclusion, for Earth has "only" warmed up by about one degree Fahrenheit (0.6 degrees Celsius) in the past century, and most analysts simply assumed that it would be impossible to detect a correlation between such a "small" amount of warming and the behavior of plants and animals. But the worldwide data used in the statistical analysis showed a clear association.

Terry's findings were controversial because they were new and surprising, and they had the potential for political influence, since people around the world care deeply about plants and animals. Loss of biodiversity is often more "real" to people than some of the other effects of global warming. Thus, some of those at the IPCC meeting who opposed stringent climate policies wanted to do everything possible to minimize the importance of Terry's findings.

Terry and I backed up this research with hard and sometimes not-so-hard data for eight hours, stretched over two days of the plenary. Despite our efforts to explain the data clearly, some delegates were quite hostile. The Saudis refused at first to agree to any language suggesting a clear scientific link between global warming and the movements of plants and animals, and many others followed suit.

We reached a stalemate, which was not encouraging, considering that in the United Nations most reports must be approved by a consensus. By consensus, the UN does not mean that most should agree, but rather that all should agree. So a contact group consisting of a selected group of the disagreeing parties, a few neutral delegates, and the relevant lead authors was formed. The goal of our group was to hammer out acceptable language on these findings for the report.

In the contact group, run admirably by Canadian scientist and IPCC bureau member John Stone, we haggled face-to-face over various concepts and wording until, in the wee hours of the morning, we were eventually able to formulate compromise language. Our two paragraphs of text conveyed the clear evidence that temperature trends were having an impact on plant and animal communities and included explicit statements on the lower level of confidence we had about how much of the behavior could be linked to human-induced global warming. We could claim with high confidence that the behaviors and movements of plants and animals that had exhibited changes were connected to observed climate change, but whether that change was due to human activities would require considerably more sophisticated analysis.

The contact group meeting ended close to 3 a.m. We were thrilled that we had agreed on language for our section and pleased that Terry's primary finding was still largely intact.

After only a few hours of sleep, we returned to the UN building, where the computerized text of the entire IPCC document was projected onto large screens on two opposite walls. Our task that day was to review the Summary for Policymakers and reach consensus on the text. The computer operator could track in red all the changes that were made to the text, so we could all see the proposed changes on the document projected on the walls. Delegates would raise their hands or their placards if they had objections and would propose modifications, which would also be tracked in red on the screen. Sometimes a single paragraph would take hours of painstaking negotiations.

At 9 a.m. on the third day of the plenary meeting, our contact group's two paragraphs, which had taken two days to negotiate, were flashed onto the screens. I held my breath, to see who would complain first. The chairperson looked around, called for comments, and quickly said, "Seeing none." He slammed his gavel down on the table and accepted the text. I could feel the adrenaline draining from my system.

While I was heartened by this initial victory for sound science, we were still faced with a serious problem. Two days of the plenary had been spent wrangling over the first major conclusion in the IPCC document, and we still had four or five more conclusions to cover. Only three days remained, and the delegates would go home at the end of the fifth day whether the document was approved or not.

But luck was on our side, and we got back on schedule. IPCC co-chair Jim McCarthy from Harvard and Bob Watson ran a tight ship. Unfortunately, the Russians, Saudis, Chinese, and a few others seemed to have saved all their complaints for the last day. They made objection after objection to topics that didn't strike me as particularly controversial. Before we knew it, it was 11 p.m., and the document wasn't finished. The meeting was set to end at midnight, at which point the translators were free to go, as stipulated in their contracts. Had these delegates intended all along to kill the document at the eleventh hour?

Bob Watson, the head of the IPCC, left the podium and was seen negotiating quietly but feverishly with the defiant delegations, especially Russia's Yuri Izrael and Mohammed Al-Saban from Saudi Arabia. Under Bob's strong leadership, the delegates slowly negotiated changes and accepted the modified text, but by that time the witching hour was ten minutes away, and there were still two paragraphs left to approve. Were we going to make it? At two minutes to midnight, the Russian head delegate interrupted with yet more inexplicable complaints and requests for trivial changes. Other countries objected to the Russians' requests. We were going to fail. Was all this work for nothing?

Bob saved the day for the second time. He looked up and said, "Translators, three years' worth of work is on the line. Can you please stay two more hours and let us complete this document?" He knew that two or three delegations—the very same ones who were completely capable of understanding English in the contact groups—had in the past refused to let the meeting proceed once the translators left, claiming it was a violation of procedure. The translators reluctantly agreed to stay, but at 2 a.m. they'd had enough.

We still had a few sentences left. Finally, the Russian delegate said, "I am putting this meeting under protest, but I will allow it to continue without translation." A Chinese delegate gave a five-minute speech in Chinese (for which no translation was available) to prove his point. Amazingly enough, after all that consternation, in a matter of 15 minutes the document was agreed upon, gaveled, and closed. A dog-tired, stressed-out group of scientists and delegates dragged themselves out of the UN building at 3:15 a.m. with a completed and approved text. The three-year process of producing the IPCC Third Assessment Report was over.

Terry and I returned home to face a very different war—my battle with mantle cell lymphoma. Over the next couple of years, I used every bit of risk assessment strategy, decision analysis, and unconventional treatment protocols I could conceive of to win out over the rare cancer that had attacked my body. That's how I was able to survive to participate in the IPCC Fourth Assessment, the Nobel Prize–winning one, and continue the battle to stabilize the planet's climate.[13]

CONSEQUENCES BECOME HARDER TO IGNORE

Weather disasters continued to worsen during the late 1990s and early 2000s. The Kyoto Protocol had not become international law yet, and few countries were taking the possible consequences of greenhouse gases and global warming seriously enough to do much self-regulating. The Montreal Protocol had halted the destruction of the ozone layer, proving that a concerted international effort can be successful. But the world was still not in action mode, even after the IPCC Third Assessment Report published its statement in 2001: "There is new and stronger evidence that most of the warming observed over the last 50 years is attributable to human activities."

Then the brown clouds gathered. A giant, smoggy atmospheric cloud made up of aerosol particles of fossil fuel and biomass emissions and other polluting chemicals and gases accumulated over South Asia and the Indian Ocean. During 2002 and 2003 the cloud drifted high above South, Southeast, and East Asia, and by December 2004, NASA scientists announced that the brown cloud had intercontinental reach, creating effects around the world.

The atmospheric brown clouds may have been caused by the rapidly emerging development in China, where cities and factories and fossil fuel–burning power plants, as well as automobiles, were being constructed at breakneck pace. Some scientists add to that much of the huge population of China who still live in rural conditions, burning biomass such as wood and dung in open kitchen stoves and cooking fires. In August 2002, the UN Environment Programme and the Center for Clouds, Chemistry and Climate released the first comprehensive report on the environmental impact of South Asian haze. But in its preliminary findings, the researchers could not determine with high confidence the relative contribution of biomass and fossil fuel burning to the brown cloud.

What the report did find, though, was that the haze was concentrated three kilometers above the surface and could travel halfway around the globe in less than a week. The cloud caused significant reduction of solar radiation to the surface, by as much as 15 percent, enough to alter regional monsoon patterns. Less sea evaporation from sunlight could mean as much as a 40 percent reduction in rain in northwest India, Pakistan, Afghanistan, and western China. More rain and flooding occurred in other areas, along with a drop in agricultural productivity and an increase in respiratory ailments.[14]

The brown clouds in 2008 were said by the UN Environment Programme to be responsible for glaciers in ranges like the Himalaya melting even faster than from global warming alone, and weather systems becoming more extreme. As China has surpassed the United States as the world's number one emitter of greenhouse gases, the implementation of international agreements on reductions becomes even more critical. Clouds are not supposed to be brown.

Extreme weather extended beyond the floods in China and the droughts in Africa to wreak havoc in Europe in the summer of 2003. A massive heat wave killed an estimated 55,000 people in Europe, most of them elderly and living in crowded cities in buildings that lacked air-conditioning or elevators. They were not designed for a week with 38-degree Celsius (100 degree Farenheit) temperatures—for good reason. It had never happened before. The severe heat wave began in June in Europe and continued through July into August. Temperatures were on average several degrees higher than the seasonal average. France recorded temperatures soaring to 40 degrees Celsius (104 degrees Fahrenheit.) The cities of Europe were experiencing a typical summer in Rio de Janeiro—but neither the buildings nor the humans were adapted to it.

August 2003 was the warmest August on record in the Northern Hemisphere; in Belgium, temperatures were higher than any in the Royal Meteorological Society's register dating back to 1833. A heat wave lasting so many weeks and at such elevated temperatures had enormous adverse social, economic, and environmental effects. Heat waves are deadly, mostly affecting the elderly, the very young, and the chronically ill.

Nearly 15,000 people died in France, the largest total for any country. Germany lost over 7,000 people to heat-related illness, and Italy, Spain, Portugal, the Netherlands, the United Kingdom, and other countries had fatalities in the thousands. The greatest losses were experienced in cities, where the heat-absorbing dark roofs and pavement increased the temperatures and inhibited the cooling temperatures at night, becoming, in effect, "heat islands." The human body simply cannot tolerate an internal temperature above 40 degrees Celsius (104 degrees Fahrenheit); the organs begin to shut down, resulting in death.

The heat wave was a wake-up call to the European Union—one more warning of impacts from a warmer climate on populations and ecosystem. Forest fires, droughts, and even the melting of glaciers in Switzerland brought the issue of climate change to the full attention of Europe's citizens. From that point on, Europe jumped onto the climate change policy wagon.

That left one significant doubter. The United States of America still resisted. If America was not willing to participate, global warming would continue largely unchecked, since the world's largest economy and historically its largest accumulated emitter would have to be part of an international effort to fix the problem, if meaningful fixes were to be achieved.

AWARENESS DAWNS ON PLANET EARTH

6 IF GLOBAL WARMING is sending out signals to beware of its wrath, the United States received its klaxon call with Hurricane Katrina. The inadequate, highly criticized response by the Bush Administration was only one, however, in a series of embarrassments as the world increasingly demanded that its governments take action. The fourth IPCC session would produce even more dire assessments, and despite fierce opposition, scientists demanded that their results be committed to public record. Unless we consider the leadership from Al Gore and his game-changing *An Inconvenient Truth*—U.S. leadership on the issue would have to wait for a change in U.S. administrations.

On August 29, 2005, Hurricane Katrina struck the Gulf Coast, followed a month later by Hurricane Rita. These were two of the most intense hurricanes ever recorded in the nation's history. The storms had a massive physical impact, affecting 90,000 square miles—an area the size of Great Britain. Over 80 percent of the city of New Orleans flooded when 53 levees were breached. More than 1.5 million people were directly affected—Katrina was the largest natural disaster ever experienced in the United States. At least 1,800 people lost their lives and more than 800,000 were forced to leave their homes. Damage was estimated at over $80 billion, not counting continuing expenses in places where refugees ended up, requiring local municipalities to care for them.

Hurricane Katrina should serve as an omen of what increasingly lies in our future, as the planet warms and the sea levels rise. Not just the United States but all the nations of the world must prepare for the likelihood of extreme weather events that will occur with unexpected force and at unpredictable times. The damage we've already done to the planet has exhibited a host of consequences—drought and flood extremes, fires, melting ice at unprecedented rates, species that alter accustomed patterns, and killer heat waves, among others. Of course, not all changes are negative, for instance, shorter routes across the melting Arctic Ocean for shipping industries, less respiratory illness in the warmer wintertime flu season, and some improvements in growing-season length in high latitudes. But as IPCC has shown, these benefits are overwhelmed by many more negative impacts, particularly if global warming exceeds a few degrees Celsius.

What we need to do is take steps to adapt to the changes we can't prevent and mitigate those risks we can't easily adapt to. Intensifying hurricanes from global warming was long ago—mid-1980s—predicted by Kerry Emanuel, a brilliant theoretical meteorologist at MIT, and has been roughly confirmed by recent studies showing increased trends in top wind speed hurricanes since 1970. Scientists readily admit the data are less than we'd like for clear trends, but the intensification has been observed in the North Atlantic, Pacific, and Indian Oceans. IPCC said it was "more likely than not that some of the intensification was due to warmer ocean surface temperatures driven by greenhouse-gas buildups."

The Bush Administration simply chalked up Katrina and the others to natural causes. While nature, not humans, does make hurricanes, warmer waters can intensify them, and if we have a hand in that warming, we are not wholly blameless victims.

Shortly after Katrina, I was interviewed on the Bill Maher HBO television program *Real Time* to discuss this.[1] I noted that hurricanes happen in summer and fall and not winter and spring because of warm ocean temperatures. Therefore, Maher asked me, could the increased temperature of the Gulf of Mexico as a result of global warming have put Katrina "on steroids"? He is fast on his feet as a political comic, and I replied that we should appoint him as President Bush's science adviser in the White House "because they haven't figured that out yet." It was great fun, but I went on to say that many factors influence hurricane intensity, sea surface temperature being just one of them. How global warming would affect those other factors is largely unknown. But, I noted, on occasions when conditions are right for a perfect storm, warmer waters will drive such storms with more intensity.

How much of the Katrina storm surge can we attribute to the warmer Gulf of Mexico from global warming? Impossible to be precise in any single event. My guess is somewhere between 3 inches and 3 feet (8 centimeters and 1 meter) of the storm surge that wrecked the city. In other words, from a negligible to very significant effect. But as we warm further, the IPCC says, it is "likely" hurricanes will intensify more, and more obviously.

Two of the scientists most associated with the hurricane–global warming connection later gave interviews in which they stuck to their predictions but toned down their confidence in that connection for recent events. Kerry Emanuel, the hurricane expert at MIT, talked about his new research technique, a method of computer modeling that predicts hurricane activity over the next 200 years. "The take-home message is that we've got a lot of work to do," Emanuel said in the interview. "There's still a lot of uncertainty in this problem. The bulk of the evidence is that hurricane power will go up, but in some places it will go down." According to Judith Curry of the Georgia Institute of Technology, a leading hurricane and climate scholar, the issue probably will not be resolved until better computer models are developed. "The generally emerging view seems to be that global warming may cause some increase in intensity, that this increase will develop slowly over time, and that it likely will lead to a few more Category 4 and Category 5 storms," she said.[2] No one knows how many or when.

BATTLES IN BRUSSELS

My appearance on television with Bill Maher was only one of a wide number of science-oriented reports in the media. The efforts of scientists to alert the world were having more impact. In 2004, the year before Hurricane Katrina, the IPCC Working Group II lead authors had met to begin planning for the Fourth Assessment Report, to be completed by the end of 2007. The lead authors held four international meetings, gathering in Austria, Australia, Mexico, and South Africa, and conducted our assessments with the benefit of ever improving technology.

In early April 2007, we held the plenary session in Brussels, Belgium, to approve the Summary for Policymakers (SPM) for Working Group II. At the end of the four-day meeting we were supposed to achieve consensus on Working Group II's contribution, and accept the underlying report's chapters and Technical Summary, which detailed all the scientific assessment studies that led to our conclusions. Every conclusion in the SPM needs a "line of sight" cross-reference to the parts of the underlying report chapters in which that conclusion is supported. Further, those chapters have to support the conclusions by citation to the relevant key literature—that is IPCC policy for all working groups. The meeting wound up in tense, last-minute negotiations on the podium when the fate of the report being accepted or rejected hung in the balance. The civil equivalent of hand-to-hand combat raged in the various contact groups as the scientists and diplomats often could not agree to a compromise in the larger conference arena.

At more than one point I wondered privately whether the framework of an international panel operating by consensus—meaning everyone, in UN terms—could ever be successful in today's world.

The plenary session opened on April 2 with encouraging speeches by co-chair Osvaldo Canziani from Argentina, former Belgian Prime Minister Guy Verhofstadt, and Stavros Dimas, European Commissioner for the Environment. Hong Yan of the WMO spoke, as did Renate Christ, IPCC Secretary, and Rajendra Pachauri, IPCC Chair. Pachauri emphasized the serious implications of the broad and complex changes in climate identified by Working Group I, whose report had just been accepted after a cantankerous wee-hours sessions that haggled over their SPM. New information had emerged since the Third Assessment Report, and he stressed the importance of the Working Group II report in identifying linkages between climate change and sustainable development around the world, and for the implementation of adaptation and mitigation measures. Our main focus in this group was to address the vulnerability of socioeconomic and natural systems to climate change, the negative and positive consequences of climate change, and options for adaptation.

The Working Group II report was based on the work of 174 lead authors, 222 contributing authors, and the participation of 1,183 experts as reviewers. Pachauri presented an overview of changes in our contributions since the First Assessment Report—the introduction of crosscutting issues and sectors, information on sustainable development, cost-benefit analyses and methodologies. Patchy, as he likes to be called, also commented on the new emphasis on integrated analysis, which he said had improved but needed to be further developed. This included my group's Chapter 19, "Key Vulnerabilities and the Risks of Climate Change."

Working Group II Co-Chair Martin Parry, a geographer specializing in climate impacts with the U. K. Meteorological Office and director of the Jackson Environment Institute, happily noted in his speech that the recommendations made at the end of the Third Assessment Report in 2001 had been implemented wherever possible for the Fourth Assessment Report. They included more quantified assessments; more complete regional coverage; more information about context, multiple stresses, and sustainable development; the role of adaptation and of mitigation; assessments of thresholds and nonlinearities; and more information on currently observed effects. He also mentioned that more often, empirical studies rather than models had been used and that information on the impacts of inaction had been incorporated.

Parry announced that lead author presentations, explaining the underlying text for each section, would be given as appropriate during discussions.

That's where the fun began.

Most of the sections of the Working Group II Summary for Policymakers were negotiated and resolved—or abandoned—in consensus over the planned four full days of the meeting, Monday through Thursday. The meeting was supposed to adjourn on Thursday about midnight, but so many open items remained unapproved that we continued on through the night. Adding pressure was an international media press conference scheduled at 10 a.m. Friday morning. Quite literally, the whole world was listening—and in fact reporters did not have to wait until the press conference to get their stories filed. Skype connections worked in the conference hall, and dozens of leaks to the press were happening all night long, particularly as dramatic moments unfolded.

Early on Friday morning, the text of a key paragraph came up for further discussion after not being resolved earlier in the plenary session or even the contact group. The final text was approved to read as follows, but not before an exhaustive argument, with a long caveat, followed by:

Nevertheless, the consistency between observed and modeled changes in several studies and the spatial agreement between significant regional warming and consistent impacts at the global scale is sufficient to conclude with high confidence that anthropogenic warming over the last three decades has had a discernible influence on many physical and biological systems.

The text as originally submitted stated its finding with "very high confidence." China, supported by Saudi Arabia, proposed lowering the confidence level to "high confidence." France, Austria, Belgium, the United Kingdom, the United States, Germany, Canada, and others opposed changing the level from what had been assessed by the lead authors. The United Kingdom, opposed by China, proposed using a statement noting that the impacts are "very likely," a functional equivalent in IPCC uncertainties lingo but a possible face-saving device for China.

The relevant lead authors involved, Cynthia Rosenzweig from NASA's Goddard Institute for Space Studies and Australian David Karoly, confirmed their strong support for "very high confidence" in the statement. They explained to the delegates that when independent lines of evidence, each one showing a similar outcome and each assessed as having "high confidence" in itself, are evaluated collectively, together they imply a much higher level of confidence across the spectrum of their conclusions.

When Rosenzweig and Karoly wouldn't back down on the strong scientific basis for "very high confidence," the Chinese changed tactics. They claimed that attributing climate change to human factors was not even a Working Group II issue, but belonged to Working Group I. I thought Cynthia was going to have a fit. It was midnight Thursday, and everybody was tired and frazzled. To my complete surprise, she said, "Dr. Schneider, who was a co-author with Dr. Terry Root on a similar study reported in our Chapter 1, is here and can answer that question."

"Thank you, Cynthia," I said, although I'm not sure how much I welcomed this unexpected opportunity.

I think this set of studies is directly relevant to Working Group II on impacts and vulnerability, because what they collectively show is that not only is climate change partially attributable to human activities—which indeed is a Working Group I issue—but here this conclusion is replicated by looking at the ice, plant, and animal records. Thus if we are already seeing a discernible impact of human-induced climate change in the plant and animal world, with very high confidence, for the mere 0.75 degree Celsius (1.3 degrees Fahrenheit) warming to date, just imagine the scrambling of ecosystems we can expect as we warm up at least 2 and as much as 6 degrees Celsius (3.6 to 10.8 degrees Fahrenheit) over this century, as Working Group I projects. It would be irresponsible for us not to report in Working Group II that impact and the potential vulnerability of the natural world to climate change anticipated by Working Group I. It also is specifically directed by Article 2 of the UNFCCC that ecosystems be allowed to "adapt naturally," and this result calls that into question.

The Chinese backed off dropping the section, I was happy to see. But the "very high confidence" kerfuffle remained in play.

After the lead authors went into detail on the scientific basis for the statement, the Chinese delegate insisted that scientists in his delegation disagreed with the assessment authors on scientific grounds. Supported by Saudi Arabia, China continued to oppose the "very high confidence" level. Remember, the IPCC operates on the basis of full consensus. The lead authors requested that if "very" were removed from the statement, a footnote should be inserted noting that the authors do not agree with the statement and believe it should have a "very high confidence" in their view. Having a government challenging widely employed and sound methodologies and then putting into question the scientific work of lead authors based on some eleventh-hour, national side science they never produced was unprecedented in the IPCC, and the IPCC scientists asked to record a formal protest. Later on Friday morning, the United States suggested deleting the confidence level altogether, and Japan countered by supporting the lead authors' request for a footnote. In the final text (see above) neither of those suggestions was implemented.

Sometime around dawn, the final language of the "very high confidence" brouhaha came to a final vote. In order not to break consensus by having China and the Saudis not assent, the "high confidence" was adopted. An unprecedented dramatic event ensued.

Roger Jones from Australia told the delegates that the behavior he had just witnessed from some countries was "scientific vandalism." Spontaneous applause from most of the hall erupted—an event never seen in a plenary. The Chinese looked livid but shaken by such an emotional public rebuke. It was Skyped instantaneously to the four corners of the world, as I soon learned by checking wire services online, which were already covering the story in almost real time. Then Cynthia slowly walked from the top of the chamber, where the IPCC scientists were billeted, to the front, handed a note to Patchy and, in full view of all, started to walk back up the long aisle. Martin Parry as chair then broke in. "Cynthia, are you walking out in protest?"

"I just gave a note to the chairman of the IPCC saying the lead authors object that good science was not incorporated into the report on this topic and we want it noted formally." Again, that courageous announcement generated sustained spontaneous applause while China endured yet a second public insult to their behavior. Cynthia proceeded up the stairs and left the hall, prompting the Skypers to tell their media contacts, and soon the world headlines reported "IPCC scientist walks out in protest."

Actually, Cynthia was not walking out in protest, she told me, she was spent. Her topic was now over and she needed a power nap and a shower before she returned for the press conference—presuming there even would be one. If the final report was not accepted, there would be no conclusions to report to the media. But the die had been cast. The meeting had been elevated from a page-two science story to a front-page political story all over the world, even before the final words were gaveled into the record.

By late Thursday evening, only the thorniest, postponed text issues remained to be haggled over. We had already exceeded the limit of the plenary meeting and had to keep postponing the press conference at which we were to announce the acceptance of the report. After five years' work, it appeared entirely possible that we wouldn't even have a consensus on the report to announce.

Terry was at home in California, nine hours earlier in the day. We were communicating constantly by Skype, so she could keep up with the session in real time—and it helped me stay alert. Some of us at the plenary were constantly Skyping each other from our laptops, passing urgent messages to the scientists on the podium or scouring the delegates we knew for support on a contested wording of the text.

During a break I called Terry close to 10:30 p.m. on Thursday.

The U.S. is still privately threatening to oppose the whole Chapter 19 in which all the supporting science for the "Reasons for Concern" and burning embers diagram is documented. It could get ugly, since I promised them I would tell all and expose them to public censure via the international media tomorrow if they try to get away with that pull-the-chapter stunt at the last minute. They may be the elephants in the room *during* Plenary, but after Plenary when the press is here, we are the elephants and they are the mice. I reminded them of that and asked for reasonable behavior consistent with the findings of the U.S. lead authors and not consistent with Bush Administration ideological pronouncements, which are anti-scientific. "We agree with you but we work for them" was one response I got. I said 'So act like Gandhi and quit if you don't believe in what you are told to do.' Yeah right, but my trump card is they know I can and will get international media coverage of their behavior if it is disjoint with our lead authors, and that the administration can ill afford that negative publicity these days. "Please don't force me to make this a political story," I pleaded with them, "I'd rather say the U.S. team was helpful." So I am hopeful this will all stay sub-rosa—we'll see in a few hours when Chapter 19 stuff comes up at the end.

"Good luck with it," Terry closed out.

One of the more contentious issues was an updated diagram of climate risks, known as burning embers. Although a central feature of the Third Assessment Report, it was left out of the 2007 report. The main opposition comprised officials representing the United States, China, Russia, and Saudi Arabia. Some scientists from other countries thought the diagram's bright orange gradients of levels of risk from increments of warming were too subjective. In its place the report used written descriptions of levels of risk. Because words are less powerful than a colorful, iconic chart, many from Europe, Canada, New Zealand, and small island states demanded to include it. Unfortunately, governments of the four big fossil fuel–dependent and producing nations opposed it. (The climate-risk diagram has since been published in the *Proceedings of the National Academy of Sciences,* so their suppression was only temporarily successful.)

The more than one hundred governments that direct the IPCC asked at the outset that Chapter 19 of Working Group II assess "Key Vulnerabilities and the Risks of Climate Change." We had not previously defined the term "key vulnerabilities," but there was a great deal of literature on different levels of projected damages from different levels of warming (up to about 5 degrees Celsius, or 9 degrees Fahrenheit). The damages assessed were distributed across many sectors, regions, and social groups—like the elderly, children, women, the poor, and indigenous communities.

The Chapter 19 authors first defined seven criteria to assist governments in assessing which vulnerabilities are "key." However, they did not attempt to weigh these criteria or rank them, since that would be a normative judgment and assessment reports were supposed to avoid being "policy prescriptive."[3] These criteria were:

Magnitude of impacts

Timing of impacts

Persistence and reversibility of impacts

Likelihood (estimates of uncertainty) of impacts and vulnerabilities and confidence in those estimates

Potential for adaptation

Distributional aspects of impacts and vulnerabilities (that is, equity)

"Importance" of the system(s) at risk (this last criterion was never assessed, as it is prescriptive)

An examination of the broad-based literature suggested that the "reasons for concern" assessed in the Third Assessment Report remained a valid way to aggregate risks, and thus the levels of warming that might be considered to cause a significant number and magnitude of impacts were examined. We concluded that the magnitude of temperature increases that would lead to many significant impacts had dropped since the Third Assessment Report, and thus the thresholds for which such damages might be triggered had to be lowered. The burning embers chart on p. 190, known as "Figure 1," illustrates this evolution of author judgments from the Third Report to the Fourth,[4] and includes an extension above 5 degrees Celsius (9 degrees Fahrenheit) warming, which was the highest level of warming that the IPCC authors examined.

Examples of unique and threatened systems include Arctic Sea ice; mountaintop glaciers; threatened and endangered species; coral reef communities, which would largely disappear from both excessive warming and acidification associated with such high atmospheric concentration of CO_2. In addition, many high-latitude and altitude indigenous cultures would be significantly threatened.

Examples of vulnerability to climate extremes include Asia megadelta cities subjected to rising sea levels and intensifying tropical cyclones (potentially creating hundreds of millions of environmental refugees) and valuable infrastructures, such as the London or New York underground systems. Also vulnerable are the elderly, who are often more likely to be hurt by unprecedented heat waves, and children, who are especially vulnerable to losses in food production in general. We were especially concerned about food stresses in drought-prone areas lacking sufficient irrigation systems, where heat and water stresses could increase hunger or malnutrition.

Other vulnerable groups included poor people in hot countries with little adaptive capacity; indigenous peoples; those exposed to hurricanes and living in vulnerable low-lying areas; people in semiarid Mediterranean climates in which drought and wildfires are already a problem; and elderly and children with asthma or other lung ailments, who would be particularly affected by urban air pollution or wildfire smoke plumes exacerbated by global warming.

The number and intensity of events with possibly irreversible damages escalates with warming, and events such as a many-century trend of up to 10 or so meters (33 feet) of sea level rise from the melting of the Greenland and West Antarctic ice sheets would become likely. So would damage to coral reefs and to oceanic phytoplankton with calcium carbonate shells, which are important food sources at the bottom

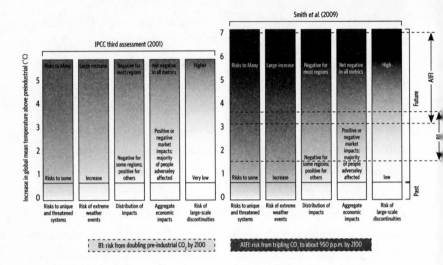

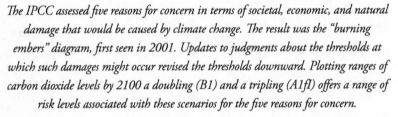

The IPCC assessed five reasons for concern in terms of societal, economic, and natural damage that would be caused by climate change. The result was the "burning embers" diagram, first seen in 2001. Updates to judgments about the thresholds at which such damages might occur revised the thresholds downward. Plotting ranges of carbon dioxide levels by 2100 a doubling (B1) and a tripling (A1fI) offers a range of risk levels associated with these scenarios for the five reasons for concern.

of the food chain, because their calcium carbonate skeletons would be vulnerable to dissolution in very acidified oceans. Tropical rain forests now relatively resistant to wildfire would become much more vulnerable. Extinction of over 40 percent of known plant and animal species would become a much more likely possibility at this level of warming (associated with doubling or more of CO_2 concentrations), particularly if climate sensitivity is in the middle to upper half of the currently estimated bell curve for it.

The revised diagram is reproduced here, without its vivid orange "burning embers." The panel on the left shows how it appeared in the Third Assessment Report, and the panel on the right shows the way we proposed to revise it based on the text in Chapter 19.

Later in the evening I Skyped Terry, in addition to what I reported above, and said: "Canada and Germany proposed adding more language from our chapter. But the U.S. is still not sure whether to try to oppose our chapter even appearing in the AR 4—that is a long story with lots of intense negotiation I had with them over sandwiches tonight. So there may yet be an ugly scene in an hour—but probably not. I think they know that if they try that, about 20 countries will angrily oppose them. And we will give the full story to Andy Revkin [*New York Times* environmental science reporter] and Seth Borenstein [Associated Press environment reporter] and tell it all in Congress next week when we testify on IPCC, so I think they will be reasonable—but who knows what the Bush Administration ordered them to do. Our delegation is probably in a tough spot. I hope they choose wisely." At nearly 1:30 a.m. Belgium time I Skyped Terry again. "You there? Things looking bleak for finishing before translators leave—Yuri threatened to make the Plenary unofficial—just like last time. Martin inexplicably declared a half hour break at 1:15 with translators leaving at 2. Fortunately delegates rejected that." We continued to battle over wording of the text, the inclusion of the figures and diagrams, and even such niggling issues as adding or removing arrows from diagrams, throughout the long night. Two a.m. came, the translators left, the usual suspects complained loudly and in native tongues about why they should protest the outcome—but like the Third Report in 2001, they behaved fine in the end after the posturing parade, and we plodded on.

Finally around 9 a.m., I got up on the podium to deal with the Chapter 19 stuff on key vulnerabilities and reasons for concern. The United States was dead set against any language referring to the UNFCCC Article—the one that warned about "dangerous anthropogenic interference" even though, as I pointed out to them, it was required of us in our Plenary Agreed Outline (PAO). They wanted almost none of our conclusions mentioned. I protested that it was central to the mission of the chapter and the report, required by the PAO, and so on. Many other countries joined in to agree. Still the United States and a few other usual suspects stonewalled.

Finally I got a Skype call from Jean-Pascal van Ypersele. "Can you hold out for five more minutes without saying much controversial? I think the U.K. is brokering a deal with the U.S." So I used up a few more minutes saying why we need to tell the world that the reasons for concern had increased and what the key vulnerabilities were.

At last Skype buzzed. "Tell Martin to call on David Warrilow of the U.K., and don't oppose this, Steve. He will get language on the reasons for concern into the paragraph in trade for not doing all the other stuff we want, but that gives us a wedge at Valencia [the next plenary meeting] for getting it all back into the Synthesis Report, as we can get a contact group on it early and not have to fight this out in Plenary at 9 a.m. the last day!"

So I scratched a note to Martin to call on the U.K., and David proposed a short sentence that the reasons for concern were still a viable way to approach aggregating vulnerability and risks. The United States and others agreed, it got gaveled down, and I kept my mouth shut, despite being angry about the integrating and original work we did in Chapter 19 at the direction of the same governments who at that moment were denying their own handiwork. But I agreed with Jean-Pascal that we now had the legal placeholder to get the camel's nose under the tent for Valencia, and there we could try to get all of it into the Synthesis Report. I was also pleased to see that I didn't have to carry out my threat to expose the U.S. delegation's behavior to the international press, since they dropped their opposition to including Chapter 19 in the report. It was less than half a loaf, but an important placeholder on the "Reasons for Concern" was secured.

The Brussels Plenary came to an end shortly after 10 in the morning on Friday, only a couple of hours before the rescheduled press conference. A last-ditch attempt by Russian delegate Yuri Izrael to preserve the "dangerous anthropogenic interference" language that we both fought so hard to keep in over U.S. objections nearly prevented the final gavel, but after Patchy spoke with Yuri, the gavel came down. We had an official report at last.

There had been fights aplenty, especially as the days had sped by with the most contentious statements postponed for further discussion. Over and over again in notes about the meeting, the phrase "Saudi Arabia, supported by the U.S.," and sometimes, Russia and China or similar groupings appeared—usually in a challenge to any specific conclusion that strengthened the assessment and thus might lead in other forums to a call for policy regarding oil or other fossil fuels. I can't say I was shocked, but I was sometimes disgusted how national interests trump planetary interests and the here-and-now overshadows long-term sustainability. I remembered my "five horsemen of the environmental apocalypse": ignorance, greed, denial, tribalism, and short-term thinking. At least three of them were riding at the Brussels Plenary.

Not all contention was ideological or special interest posturing, however. Some of the conflicts stemmed from genuine differences within the mainstream community of scientists who acknowledge anthropogenic climate change. After the final reports of Working Group I and Working Group II were gaveled into history, some contradictory statements remained about the assessment of key issues such as sea level rise. I had a major disagreement or two with my respected friend and colleague Susan Solomon, the brilliant scientist who had led the expedition to Antarctica and explained the human origins of the ozone hole. She can be pretty assertive over what she thinks is the appropriate framing of scientific issues. Working Group I, which she co-chaired, assessed the level of sea rise attributed to global warming or other causes. My chapter on impacts contained radically stronger conclusions about sea level rise than what Susan's team concluded. They had to be reconciled for the final Synthesis Report half a year after the science reports were completed. After private negotiations (often at dinner at writing team meetings, and usually over a good bottle of red wine), in the end Susan and I proposed that the Synthesis authors drop the numbers in Working Group II and simply agree to say "risk of meters of sea level rises in centuries to millennia" if it warmed up 1 to 4 degrees Celsius (1.8 to 7.2 degrees Fahrenheit) above 1990 levels. Our colleagues agreed.

Scientists enjoy these exchanges and fundamentally agree about most things, especially the need for quality and integrity in science. However, we have very different worldviews about the extent to which scientists should be engaged in public policy debates and about discussing conclusions publicly before there is a strong scientific consensus attached to them. I kept arguing that consensus should not necessarily be over conclusions, but rather the confidence we have in all important conclusions, even the less likely ones, if they imply significant impacts like meters of sea level rise. Society, not scientists, should decide how to react to more uncertain but highly significant potential risks, and we shouldn't keep it inside the club of science until we have consensus, since that is a normative risk management judgment that only society should make. We still wrangle over that issue today. The discussions between Susan and me were honest differences over issues like the definition of "consensus," and their eventual resolution through scientific data assessment—and clear airing of what was science and what was philosophy. This allowed us to enthusiastically support compromise findings of the IPCC, despite seeming contradictions from different working groups, and push forward together, as we did at the Valencia Synthesis Report Plenary.

When that meeting took place in Valencia, Spain, in November 2007, we prevailed on the issue of dangerous anthropogenic interference—and appreciated the floor help from Yuri Izrael, a vindication for him for his 25th-hour complaints on the cutting of this language from the Working Group II SPM in Brussels. We did have the one-liner compromise that Jean-Pascal and David Warrilow brokered with the U.S. on why reasons for concern were still a viable approach, allowing us to push hard for a much better summary at Synthesis. Patchy had the good management judgment—urged on by Susan and me, among others—to convene a contact group on it the very first day. In fact, he asked me to do an unofficial information session on the reasons for concern at lunch—more than half the delegates voluntarily attended. Despite the fact that several delegates from other countries told me that the U.S. delegation was initially telling other delegations that this was all a personal crusade of mine, the noon information session was a strategically good move. That contact group on the reasons for concern was set up well before the end, when the Article 2 topic came up sequentially in the report, and met twice a day all week to get a good compromise.

Also, Phil DeCola from the U.S. delegation tried hard to keep some of the senior delegates more reasonable and was a positive force in getting many good contributions from the United States. As in Brussels, they exhibited some infuriating blocking behaviors, but at other times, Ko Barrett, a senior U.S. delegate, was very helpful in providing more precise language on important points. We understood that the U.S. delegates were more personally persuaded by the mainstream science positions than by their marching orders from delegation head Harlan Watson and the Bush Administration.

The final section of the Summary for Policymakers, "The Long-term perspective," included this highlighted statement:

Determining what constitutes "dangerous anthropogenic interference with the climate system" in relation to Article 2 of the UNFCCC involves value judgments. Science can support informed decisions on this issue, including by providing criteria for judging which vulnerabilities might be labeled "key."

Key vulnerabilities may be associated with many climate-sensitive systems, including food supply, infrastructure, health, water resources, coastal systems, ecosystems, global biogeochemical cycles, ice sheets and modes of oceanic and atmospheric circulation.

You can bet that the Fifth Assessment Report (AR5) due to be completed before 2014 is going to contain the phrase "dangerous anthropogenic interference with the climate system." It will also contain, if our Scoping meeting in Venice in July 2009 holds through several plenary events, much more assessment of regional impacts, vulnerabilities, and adaptation strategies, more chapters on adaptation potential and obstacles, more coordination across working groups to avoid events like the sea level differences in the Fourth Assessment Report, and more on uncertain probability but high-consequence events like Greenland passing a "tipping point" and irreversibly melting with meters of sea level rise as a consequence. It was interesting in Venice that economists and ecologists in Working Groups III and II were often more supportive of looking at these outlier high-risk events than the climate physicists in Working Group I. Needless to say, I strongly urged my climate science colleagues to try to assess these potentially big changes, but to use confidence language that realistically assesses how little we may now know about the likelihood of such consequential possibilities. It was obvious there was considerable reluctance from many Working Group I scientists to peer into the shadowy tails of the dangerous end of the probability distribution, but as I told them several times, "If you don't do it, then those less well qualified will likely do it for you, as it is essential to the risk-management framing of the problem that governments will insist that Working Groups II and III adopt. I am hopeful that this time we will have more cross-working group cooperation and coordination far enough in advance to get differences resolved before the Synthesis Report, and not during it as in the Fourth Assessment. Time will tell.

A STORY WITH A HAPPY ENDING

The IPCC Fourth Assessment Report was far from my only engagement in the climate wars during 2007. Living as I do in California, I was very concerned with the Bush Administration's refusal to allow California to enact standards for automobile emissions and fuel mileage that exceeded those of the federal government. Vehicles account for one-fifth of all CO_2 emissions in the United States and over 40 percent in California.

Back in 1991, I had testified in congressional hearings on Corporate Average Fuel Economy (CAFE) standards under consideration by Congress at that time, which were violently opposed by the automobile industry. They wanted to keep churning out their gas-guzzling, pollution-emitting all-American cars and trucks.

A further effort in 2007 for an EPA waiver of preemption against California for its tailpipe emissions standards law, which I fought for in Washington with California state senator Fran Pavley—the author of the law—and former Governor Jerry Brown, was unsuccessful. The EPA refused to grant the waiver. The Bush Administration made no efforts to insist that the automobile industry clean up its act voluntarily or by regulations.

In April 2009, scientists Mark Z. Jacobson—a Stanford colleague in civil and environmental engineering—and Michael Kleeman, Ben Santer, Jim Hansen, and I, represented by the Environmental Law and Justice Clinic at Golden Gate University, submitted comments in response to the EPA's notice soliciting public comment on whether "there is merit to [EPA] reconsidering its [March 6, 2008] decision denying California's waiver" under section 209 of the Clean Air Act. We requested that the EPA reverse its decision and grant California's petition for a waiver. We wanted to provide the EPA with accurate information about climate change and its impacts because the agency's decision to deny California's request for a waiver has grave implications for Earth's climate and climate change impacts, including those unique to California.

Developments in climate change science underscore the critical need to reduce greenhouse gas (GHG) emissions now. Without immediate reductions, Earth may cross several tipping points where the level of GHGs in our atmosphere causes the climate system to reach a threshold, committing irreversibly to abrupt and/or major changes. Some of the major changes could become unstoppable and, for the foreseeable future, difficult or impossible to reverse.

EPA Administrator Stephen Johnson had no credible scientific evidence for his conclusion that California suffers generalized, not unique, impacts of climate change. In fact it was reported that most EPA scientists had recommended approving the California waiver, and only after the White House called Johnson did the EPA administrator change his mind. In fact, testimony to the Senate by Jason Burnett from the EPA (and a former Stanford student in my honors seminar), who watched this from the inside, said that the administrator made a singular decision to reverse the agency's scientific advice under White House pressure—something Johnson had, under oath earlier, specifically denied he received.

Climate change disproportionately affects California's air quality and hydrology. By increasing air pollution, including tropospheric ozone, anthropogenic warming kills Californians at a rate higher than in the rest of the United States. Human-induced changes in the hydrology of the western United States also impact California disproportionately. These impacts amount to "compelling and extraordinary" circumstances, legal criteria that should have led the EPA to grant California's request for a waiver. The EPA's decision to deny California's waiver ignored science and had grave, negative implications for stabilizing Earth's climate.

Fortunately, one of the first actions that Barack Obama took on assuming his office in January 2009 was to direct the EPA to reconsider the Bush Administration's denial of the California waiver. He also instructed the Department of Transportation to come up with rules to enforce a 2007 law that required a 40 percent improvement in gas mileage for cars and light trucks by 2020. That law was completely toothless when a Texas oilman occupied the White House.

The happy ending? After years of stonewalling California on its sensible limits to automobile pollution, the federal government finally enacted a sweeping change. Not only did the Obama-led EPA approve California's waiver of preemption on June 30, 2009, but it was also expanded so that the entire country would follow California's laws, cutting gas consumption by 40 percent by 2016. Although cautious, the auto industry said it would not legally challenge the ruling. Perhaps some of that cooperation followed from the bailout of Detroit that the Obama Administration was implementing, but perhaps it signaled a new recognition of the need to produce efficient cars. At last, a substantive step forward from the U.S. government in saving our planet's climate—and maybe the auto industry, too, by encouraging competitive efficient vehicles—would be made. It shows that bipartisan leadership—whether from Fran Pavley, Arnold Schwarzenegger, or Barack Obama—really matters in getting things done well.

WHAT PRIZE?

At home in Palo Alto, the phone rang early in the morning on October 12, 2007. I had just returned from Wyoming, where I'd been arguing vociferously with the coal industry about CO_2 emissions and climate policy, and I was sleeping it off.

Terry answered the call. Wayne Freedman, a San Francisco TV news reporter wanted to interview her that morning at the "press conference." Terry was interviewed all the time and several times by Wayne, so she wasn't surprised at first.

"But what press conference?"

Wayne chuckled at her apparent sense of humor. "And congratulations on the prize, too," he added.

"What prize?"

"You can't mean they didn't call you—"

"Who didn't call, Wayne?" Then he told her that the IPCC had just been co-awarded the Nobel Peace Prize with Al Gore.

"But there are a thousand of us. Who would they call first? Okay, tell me where it is and we'll see you there. Thanks."

Terry rushed into the bedroom and woke me up. "Get up and get into a suit, we're going to a press conference!"

"What press conference?" I asked.

She gathered as many local lead authors of the IPCC Fourth Assessment Report as she could find, and we made a mad dash to Palo Alto, where the press conference was going to be held. Al Gore, coincidentally, was speaking there that morning at the Alliance for Climate Protection—a new environmental group he personally funded. When the Nobel Prize was announced, a hasty press conference was set up, conveniently—and coincidentally—for us, in our home base.

I was apprehensive about crashing Al's party, but then the IPCC *was* a co-recipient. And since I had testified at one of his first climate hearings in 1981, I was sure he wouldn't mind. "Can you call his staff?" Terry asked while we were driving.

"No," I said, "I always e-mail them."

"Oh well, we'll just hope they're okay with it."

As we were struggling for parking in this now very crowded spot, my cell phone rang—it was Kalee Krider with the Gore team. "Just called your office and I got this number. Want to come to the Nobel press conference?"

Terry and I and our lead author colleagues Chris Field and Thomas Heller joined Al and Tipper Gore on the podium in front of the cameras. I was disappointed that my former student and now research colleague Michael Mastrandrea, who was also an IPCC author from Stanford, was out of town that day. It would have been the thrill of a lifetime for Mike. It was a great day for the climate change campaign. After a five-minute statement to the army of cameras and print reporters, the Vice President ended, "The IPCC lead authors will answer your questions." As he walked off, almost all of the cameras folded up and left, and maybe ten total reporters—including Wayne—remained to hear what the four of us had to say. I immediately realized why Gore had departed: The reporters only wanted to ask Gore about whether he was running for President or whom he would endorse—Obama, as it later turned out. Al wanted climate, not him, to be the star of the day.

This is why I was so pleased about the IPCC part of the prize. The cameras came primarily because of the cult of personality, but IPCC represented the culture of community. We can't assess complex systems science individually, nor can we solve the global policy problem without coalitions and communities with common purpose. Al Gore understood that when he walked off, and soon most of the rest of us will too. It was nice to have had this event, I thought to myself, a dramatic contrast to my more typical interactions with some power people. For example, around 2007, Senator James Inhofe (Republican of Oklahoma), the most aggressive anti-global-warming voice in the Senate, read a statement into the Congressional Record saying that I was the father of the greatest environmental hoax. I recall sending some e-mail to his office thanking the senator for the honor, but respectfully declining as I have a thousand equally deserving colleagues. But with the Nobel announcement and press event, I decided we hadn't lost the battle yet, though more struggle was certain to come.

THE MEDIA WARS:
THE STORIES BEHIND PERSISTENT DISTORTION

7 **THE TACTIC OF PERSISTENT** distortion is nothing new in the battle arena of climate change. Over the years my colleagues and I have been the targets of personal attacks and subject to false reporting, biased interpretations paid for by lobbies and big business, and other violations of media ethics. And the problem has only gotten worse.

One of the key reasons for distortion in the media reports on climate change is the perceived need for "balance" in journalism. In reporting political, legal, or other advocacy-dominated stories, it is appropriate for journalists to report both sides of an issue. Got the Democratic view? Better get the Republican.

In science, the situation is radically different. There are rarely just two polar-opposite sides, but rather a spectrum of potential outcomes, which are often accompanied by a history of scientific assessment of the relative credibility of each possibility.

A climate scientist faced with a reporter risks getting his or her views stuffed into one of two boxed storylines: "We're worried" or "It will all be okay." Sometimes, these two boxes are misrepresentative: Amainstream, well-established consensus of hundreds of experts may be "balanced" against the opposing views of a few special inter-est Ph.D.'s. To the uninformed, each position seems equally credible. This is particularly true when the issues are complex and most normal people can't discern which side is more credible in a verbal duel using technical terms. Indeed, one strategy of deniers is to raise ten points of objection in their two minutes in a debate. The defender of the main-stream views can't possibly address all ten issues—at best, he or she can meaningfully answer two or three. By that tactic the challenger, no matter how lacking in credibility among the cognoscenti, achieves equal status by gaming the system.

The way I handle that situation is to show why two of the issues on the laundry list of challenges are demonstrably wrong and then explain to the audience why the format is unworkable. Perhaps that helps a lit-tle, but a well-moderated debate, like a presidential confrontation where the moderator tries to keep it focused on one issue at a time, should allow only one or two points at a time to be brought up for discussion. That way, they can be answered sufficiently to show who is more likely to be right—although even that is still a chore for most folks when the topic is technically dense. Today's scientists must learn quickly how the advocacy system and its media accompanists really function.

Being stereotyped as the "pro" advocate versus the "con" advocate as far as action on climate change is concerned is not a quick ticket to a healthy scientific reputation as an objective interpreter of the science—particularly for a controversial science like climate change, which is rarely one-sided. In actuality, it encourages personal attacks and distortions. This is all part of the problem I call, somewhat whim-sically, "mediarology."

The critical importance of some issues—such as policies to make our air healthier and our waters cleaner and our sea levels more sta-ble—requires policies that involve public support for the courageous politicians willing to take them on. Public support, in turn, is built on at least a partial knowledge of the issues—even for complex topics like climate change. If the public is bamboozled into thinking that each "side" in a typical media debate is credible, a typical reaction is to say, "Well, if the experts don't know, how can I know! Let's just wait a while until they figure it out." That is precisely the strategy of the climate deniers, to create public confusion and apathy, which slows policies that help the planet and our children but hurt the special interests who are vested in the status quo. In this sense mediarology is a topic we all need to recognize, so we are better informed and thus are more willing to express our support for protecting the ozone layer, clean water, and the stability of the climate.

When I give a public talk on aspects of climate change, I always take the time to explain the difference between climate deniers and skeptics. All good scientists are skeptics—we should challenge everything. I was a big-time climate skeptic, changing from cooling to warming and nuclear winter to nuclear fall when that is where the preponderance of available evidence led. As more solid evidence of anthropogenic global warming accumulates, the numbers of such legitimate climate skeptics are declining. Climate deniers, however, are not true skeptics, but simply ignore the preponderance of evidence presented. Skeptics should question everything but not deny where the preponderance of evidence leads. The latter is, at best, bad science or, at worst, dishonesty.

COURTROOM EPISTEMOLOGY

Expert witnesses spouting opposing views—in Congress, courtrooms, or on editorial pages—often obscure an issue more than they enlighten. They refuse to acknowledge that the issue is multifaceted and present only their own argument, ignoring—or denigrating—opposing views, no matter how credible they are to the mainstream community of experts. This is no big surprise, but what is shocking is how often that strategy is deliberate. Stakeholders increasingly select information out of context to protect their interests (ideological or financial), and clear exposition and assessment in context have sunk even lower on their priority lists.

In my 2007 testimony to the U.S. House of Representatives Science and Technology Committee, Congressman Dana Rohrabacher from Orange County, California—a long-established climate contrarian—tried to challenge the testimony of four IPCC Working Group II lead authors by noting that temperatures since 1998 had not risen and thus he had proven global warming to be false. I tried—several times—to explain that climate involves trends extending over many decades. Had I showed a record from 1992 to 2000, it would have looked like we were going to hell in a hand basket, since Earth was warming so fast. All such short-term runs have little to do with climate trends. It would be like trying to estimate Willie Mays' lifetime batting average from how he did in August 1960. It would be nothing more than a statistical fluke. At the hearing I did not convince the congressman, but I was gratified the rest of the members apparently understood what a sampling error is, and the polemics stopped there—although Rohrabacher continues to spout this nonsense frequently.

But in the popular media Rohrabacher's testimony would have been a sound bite like "It hasn't warmed since 1998 though greenhouse gases have increased, so global warming is refuted," followed by mine saying, "It is too short a time to see any significant trend and what happened since 1998 proves nothing." An audience of nonexperts could come away pretty confused and not ready to endorse strong measures to solve a problem about which they still had big doubts.

The attitude typical of legal-eagle antics says, "It's not my job to make my opponent's case!" and arises not only in courtroom histrionics, but also in most political debates and in much of the media. I call this "courtroom epistemology." The snippet of fact disembodied from fair context by Congressman Rohrabacher is a typical instance. Another example comes from the deniers who point out that the snow on the top of Greenland is building up and thus there is no threat of sea level rise from deglaciation of its massive ice sheet. Meanwhile, they neglect to mention that the sides of Greenland, which are warmer, are melting much more ice than is accumulating on the top—and thus a sea level rise is indeed a big risk. Moreover, this interpretation neglects the prediction from climate models that ice will build up on top of the big ice sheets until the atmospheric warming is so large that the temperatures rise above freezing up there, which is already happening on the flanks of Greenland and creeping up the ice to the interior. So instead of refuting the model's predictions, as contrarians insist, the buildup of ice on the top and the melting of the sides actually validates the predictions. Yet the courtroom epistemologists trying to raise doubts only mention the ice buildup on top and conveniently forget the rest of the story.

It's disgraceful that the media allows such routine distortions in complex system debates like climate change, as if a fact is somehow an "opinion" and all opinions should be aired. If the opinion were that the writer doesn't think the net melting is important enough to build policies to hedge against it—fine, that is an opinion and belongs in the op-ed space. But to allow known falsehoods or misframings of science is not an opinion, just an error or worse. That should in my view be distinguished from real opinions—value judgments on what we should do about it, for example—and a newspaper has a right to demand that such demonstrable factual errors be removed. If a political writer claimed blacks were better off in the Jim Crow South than now, would that be an "opinion" they would publish in their newspaper? Or that smoking doesn't cause cancer? You get the point.

The question isn't whether reporters, politicians, lawyers, and others or their methods are wrong or that "impartial" scientists are morally superior—but whether the techniques of advocacy-as-usual are suited for a subject like climate change in the public arena. In the advocacy arena, everybody knows the game—spin for the client. But in science, the playing field for public discussions is not level. Any spin on the facts would cause damage to a scientist's reputation—especially young scientists. That is decidedly not true for a status quo defender advocating for client interest. They are rewarded for winning, not for fairly reporting evidence.

Scientists think that advocacy based on a "win for the client" mentality that deliberately selects facts out of context is highly unethical. Unaware of how the advocacy game is played outside the culture of scientific peer review, scientists can stumble into the pitfall of being labeled as advocates lobbying for a special interest, even if they had no such intention.

When a scientist merely acknowledges the credibility of some disputed information, opposing advocates often presume the expert (scientist) is spinning the information for some client's benefit. The Ben Santer story from Chapter 5 is a case in point. Even when the scientist points out that there is a wide range of possibilities and refers to extensive peer-reviewed assessments, the opposition accuses the expert of currying favor from some alleged funding agent. After all, isn't that what everybody else is doing?

The late best selling novelist Michael Crichton made exactly that claim in an interview with ABC News *20/20*'s anti-global warming co-anchor John Stossel in December 2004. Stossel, discussing Crichton's new novel, *State of Fear,* pointed out that the book assumes people studying global warming have an incentive to exaggerate the problem to get grant money. "Everybody gets their grant by doing that," Crichton said.

"And if you say, 'there isn't a big problem,' you're less likely to get money?" Stossel asked.

"Absolutely."[1]

On the contrary, if scientists really exaggerated the risks of climate change, society would no longer need to fund us because our job would be over. The best way for us to curry funding favor would be to deny we know much and stress uncertainties to get research money to reduce them. Overstating uncertainty, not exaggerating risks, would be a much better trick for that dishonest goal we are accused of pursuing by many in the climate denier camp. They couldn't even get their smear logically right.

The fundamental question related to climate change and mediarology is: How can we assist, or at least encourage, advocates to convey a balanced perspective when the "judge" and "jury" are Congress or public opinion, and the "lawyers" are the media-wise advocates and the media itself? The polarized advocates get only ten-second sound bites each on the evening news or five minutes in front of a congressional hearing to summarize a topic that requires hours just to outline the range of possible outcomes, much less convey the relative credibility of each claim and rebuttal.

But scientists, like everybody else, have unconscious biases too. The more scientists discuss our initial assessments with colleagues of various backgrounds, the higher the likelihood we can illuminate unconscious biases. We may not ever reach the archetype of pure objectivity—but pure objectivity is, of course, a myth in science. The path to objectivity does not involve scientists holding back their opinions in order to maintain a pretense of some higher calling as objective scientist. Rather, only active effort to make our biases conscious and explicit via outside review is likely to keep our science-advocacy more objective.

What about the actual legal battles that take place over climate change in U.S. courts, where judges try to decide on complex science issues? A recent case in point illustrates why courts are not the best places to make science into law. In a *New York Times* Dot Earth blog about the EPA's new classification of carbon dioxide as a pollutant under the Clean Air Act in April 2009, environmental science reporter Andrew Revkin credits his colleague John Broder with saying that it's likely that the "pollutant" label is intended to add "pressure to congressional efforts to write new gas-limiting legislation." Some of that legislation will be challenged in the courts, for sure. We can only hope that the U.S. Supreme Court justices will have boned up on their science knowledge by then, to avoid interchanges such as the one that took place in November 2006, recounted by Revkin. During Supreme Court arguments over whether heat-trapping gases harmed the environment and public health, confusion arose over which layer of the atmosphere trapped heat from the CO_2 emissions from tailpipes and smokestacks. The assistant attorney general of Massachusetts corrected Justice Antonin Scalia, saying, "Respectfully, Your Honor, it's not the stratosphere. It's the troposphere."

"Troposphere, whatever," Justice Scalia replied. "I told you before I'm not a scientist. That's why I don't want to have to deal with global warming, to tell you the truth."[2] In *Massachusetts* v. *E.P.A.* the Court, by a razor-thin 5-4 vote, ordered the agency to undertake studies that ultimately paved the way for the "pollutant" designation once the Bush Administration could no longer suppress the work. The Obama EPA is doing just that, but expect more court actions and delaying tactics as they proceed. (Incidentally, it is *both* the stratosphere and troposphere that contribute to CO_2 heat trapping.)

At least some encouraging signs are emerging. In July 2008, a vociferous debate ensued when Andrew Revkin published a thought-provoking article titled "Climate Experts Tussle over Details. Public Gets Whiplash." Revkin explored the blow-by-blow media coverage of scientific findings on global warming issues, which is a different issue from the "false balance" issue just described. Scientists and contrarians immediately rushed to Revkin's Dot Earth blog on "Whiplash Effect and Greenhouse Effect" to defend or deny.

Revkin introduced the controversy over news reporting of conflict-
ing science in carefully chosen words:

> When science is testing new ideas, the result is often a two-
> papers-forward-one-paper-back intellectual tussle among compet-
> ing research teams.
>
> When the work touches on issues that worry the public, affect
> the economy or polarize politics, the news media and advocates
> of all stripes dive in. Under nonstop scrutiny, conflicting find-
> ings can make news coverage veer from one extreme to another,
> resulting in a kind of journalistic whiplash for the public.
>
> This has been true for decades in health coverage. But lately
> the phenomenon has been glaringly apparent on the global
> warming beat.
>
> Discordant findings have come in quick succession. How fast
> is Greenland shedding ice? Did human-caused warming wipe
> out frogs in the American tropics? Has warming strengthened
> hurricanes? Have the oceans stopped warming? These questions
> endure even as the basic theory of a rising human influence on
> climate has steadily solidified: accumulating greenhouse gases
> will warm the world, erode ice sheets, raise seas and have big
> impacts on biology and human affairs.
>
> Scientists see persistent disputes as the normal stuttering jour-
> ney toward improved understanding of how the world works.
> But many fear that the herky-jerky trajectory is distracting the
> public from the undisputed basics and blocking change.[3]

Before he wrote the *New York Times* piece, he e-mailed a group of
scientists of differing perspectives and asked us what we thought. A
lively debate ensued with many opinions, several of which appeared
in the eventual article published. Revkin often does this, generating a

private debate among many that he can study to help his writing be truly balanced—reporting on the spectrum of views, not just giving equal weight to two extreme opposing radicals. I was personally gratified that he thought my e-mails on this were worth making it to the published *Times* story:

> To support clarity, Stephen H. Schneider, a climatologist at Stanford, helped create a glossary defining what is meant by phrases like "very likely" (greater than 90 percent confidence) in the reports from the Intergovernmental Panel on Climate Change. In a news media universe where specialized reporting is declining and a Web mash-up of instant opinion and information is emerging, Dr. Schneider said, it is ever more important for scientists to take responsibility for communicating in ways that stick, while sticking with the facts.[4]

THE "DOUBLE ETHICAL BIND" PITFALL

Would you trust a scientist who advises his or her colleagues to use scary scenarios to get media attention and to shape public opinion by making intentionally dramatic, overblown statements? Would you have confidence in his or her statements if the scientist said that "each of us has to decide what the right balance is between being effective and being honest"? Understandably, you'd probably be suspicious and wonder what was being compromised.

I confess: Those were *some* of my words, yet their meaning is completely distorted when viewed out of context like this. In hundreds of places—especially on the websites advocating industrial or economic growth opposed to global warming policies that might harm their or their clients' interests—I am similarly (mis)quoted, alongside a declaration that my environmental cronies and I should never be trusted. It is an old trick: Blame the messenger, even with misquotes.

This example illustrates the risks of stepping from the academic cloister out into the wide world. A scientist's likelihood of having his or her meaning turned on its head is pretty high—especially with highly politicized topics such as global warming. First, consider a movie theater marquee selectively quoting a critic as having said a movie was "spectacular," when the critic might have actually written, "The film could have been spectacular if only the acting wasn't so overplayed and the dialogue wasn't so trite." You get the idea.

My first major experience in being so misrepresented in the public debate occurred after the 1988 heat waves in the United States, when global warming made daily headlines. I gave a dozen interviews a day for several months that year. The global warming debate migrated from the ivy-covered halls of academia and bland offices of governmental scientists into the public policy spotlight. There were congressional hearings, daily media stories and broadcasts, pressure on the government from environmental groups pushing for control of CO_2 emissions—and loud and angry denial by industries and countries with high CO_2 emissions of both their contribution to global warming and the credibility of the science behind climate change.

I was frustrated about the capricious sound-bite nature of the public debate, and I expressed my frustration to Jonathan Schell, a Pulitzer Prize–winning writer doing a story on the contentious climate debate for *Discover* magazine. I guess my first mistake was to be a bit tongue-in-cheek—I painted a stark picture of the opposing viewpoints in the climate change debate: gloom-and-doom stories from deep ecology groups and others versus pontifications on uncertainties from big industry and allies, who used that to argue against preemptive action. I complained that even though I always make a point in my interviews to discuss the wide range of possibilities, from catastrophic to beneficial, media stories rarely convey the entire range.

I tried to explain to Schell how to be *both* effective and honest: by using metaphors that simultaneously convey both urgency and uncertainty, and also by producing supporting documents of all types and lengths, from op-eds to full-length books. Unfortunately, this clarification is absent from the *Discover* article, and this omission opened the door for 20 years of subsequent distortions and attacks. Ironically, this is the consummate example of my grievance about problems arising from abbreviated versions of long interviews.

Here is the published quote from that interview with *Discover,* from which selected lines have been used for over a decade as "proof" that I exaggerate environmental threats:

On the one hand, as scientists we are ethically bound to the scientific method, in effect promising to tell the truth, the whole truth, and nothing but—which means that we must include all doubts, the caveats, the ifs, ands and buts. On the other hand, we are not just scientists but human beings as well. And like most people we'd like to see the world a better place, which in this context translates into our working to reduce the risk of potentially disastrous climate change. To do that we need to get some broad based support, to capture the public's imagination. That, of course, means getting loads of media coverage. So we have to offer up scary scenarios, make simplified, dramatic statements, and make little mention of any doubts we might have. This "double ethical bind" we frequently find ourselves in cannot be solved by any formula. Each of us has to decide what the right balance is between being effective and being honest. I hope that means being both.[5]

The *Detroit News* selectively quoted this passage, already not in its full context, in an attack editorial on November 22, 1989: "On the one hand, as scientists we are ethically bound to the scientific method. On the other hand, we are not just scientists but human beings as well. To do that we need to get some broad based support, to capture the public's imagination. That, of course, means getting loads of media coverage. So we have to offer up scary scenarios, make simplified, dramatic statements, and make little mention of any doubts we might have. Each of us has to decide what the right balance is between being effective and being honest."[6]

Notice anything missing? The most glaring omission in the *Detroit News* quotation is of the last line of the *Discover* quote, the one about being *both* honest and effective. The *Detroit News* clearly misquotes me, presumably since including the addendum would have weakened the effectiveness of their character attack. In response, I prepared a rebuttal containing the full quote and the context of my interview, which actually showed that I disapproved of the sound-bite system and the media's polarization of the climate change debate.

While the *Detroit News* readers had an opportunity to see my true intent, albeit a month later when the rebuttal was published, I simply cannot respond and correct every article misquoting me, as they have proliferated and now number in the hundreds—and the thousands in anti-warming websites and blogs. Despite many attempts on my part—in my books, papers, talks, and other op-ed pieces—to outline my opinions and dispel the media-propagated myths, the distortions continue to this day, even in "respectable" publications like the *Economist*.

Unbelievably, in March 2009 while I was testifying to the New Zealand Select Committee on Emissions Trading, wouldn't you know it—one opponent of climate policy sure enough drags out of his box of distortions the old half quote. When I got done showing how irresponsible he was in front of his colleagues and the media, he was no doubt sorry he took that path. Yet those disinformation mills like the Competitive Enterprise Institute in Washington just churn out this stuff and the faithful use it as if it were truth—after all, most of the debaters they face won't know the full story and most readers and listeners won't search Google to see if it is true. Remember, in the advocacy world the object is victory, not truth.

In 2002 the *Economist* ran a partial quote (also taken from the *Discover* article) without even calling me to see if it was valid. Referring to my criticism of Danish statistician Bjørn Lomborg's error-riddled 2001 book, *The Skeptical Environmentalist,* the editor wrote, "The fuss over Mr. Lomborg highlights an attitude among some media-conscious scientists that militates not just against good policy but against the truth [followed by a partial quote of the *Discover* article]. . . . Save science for other scientists, in peer-reviewed journals and other sanctified places. In public, strike a balance between telling the truth and telling necessary lies. Science needs no defending from Mr. Lomborg. It may very well need defending from champions like Mr. Schneider."[7] I wonder how many readers of the *Economist* looked it up to see that the magazine was the party guilty of distortion and lack of journalistic due diligence.

The most egregious distortion on this standard lie appeared in a 1996 opinion piece by the late Julian Simon, a business professor at the University of Maryland—and a hero to Bjørn Lomborg. Simon not only used an out-of-context quote from the *Discover* article to "prove" that I advocate exaggeration in order to get attention, but he also *invented* a preamble that I advise people to "stretch the truth," and he attributed that to me.

Some friends have advised me to file lawsuits against such distortionists engaging in showcase "journalism," but as a public figure I have learned to deal with character assassination and polemics as part of the "real world" of public policy debate. Moreover, lawyer friends have told me that partial quotes, even those that turn the original meaning of the full statement upside down, are generally protected by the First Amendment. In the face of this no-win scenario, I warn those who venture into this quagmire simply to expect that most people do not check the original quotes or stories for accuracy or fairness, often wrongly thinking the media do so.

During the debate over the Kyoto Protocol, in December 1997 conservative columnist Charles Krauthammer wrote an op-ed piece in the *Washington Post* entitled "Global Warming Fundamentalists: This Is Nuclear Winter Without the Nukes." He lumped Vice President Gore and me into a stereotype of arrogant overconfidence: "Uncertainty is a feeling foreign to global warming fundamentalists.. . . Stephen Schneider, a Stanford scientist and participant at Clinton and Gore's global climate change roundtable last July, said that when it comes to global warming, it is 'journalistically irresponsible to present both sides.'[8] Of course, Krauthammer neglected to say that I favor presenting *all* credible "sides," not just two of them, as that bipolarity rarely exists in complex science. I also favor presenting an assessment of the scientific credibility of each claimed outcome.

Despite hate mail from the peace movement, I was the one who changed "nuclear winter" to "nuclear fall" in 1984 because that is how the revised science came out. How is this ends-justify-the-means arrogance? Krauthammer went on to say that "with the zeal of the convert . . . Schneider [says] that journalists shouldn't even present the non-global warming view, as does Gore when he makes the skeptics into the moral equivalent of tobacco executives."[9]

My rebuttal, published in the *Washington Post* a month later, well after Kyoto was over, put the record straight that I do not advocate leaving out any opinion but reporting relative credibility of multiple views:

To quote a hundred-scientist assessment in one sentence and then "balance" the story by giving equal space and credibility to one of a handful of contrarian scientists who represent a tiny minority of knowledgeable opinions *is* irresponsible journalism in my opinion. Such false balance projects a distortion of the mainstream knowledge base of the scientific community because it represents all opinions as somehow being equally credible, even though thousands of scientists have worked for years to sort out the likely from the unlikely—and we're still doing that because science is never 100 percent sure of anything . . . In short, I am not now and never have been in the ends-justify-the-means club.[10]

This appeared in the *Washington Post*, but the original Krauthammer op-ed piece was syndicated to God knows how many newspapers that never saw my rebuttal.

As angry and frustrated as being misrepresented makes me feel, the real rub is that many people of goodwill actually believe it. I had a poignant example of this in 1995 at lunch with the leaders of IPCC Working Group I, of which I was a lead author. When our chairperson asked me to help them with media, a senior Australian scientist said he opposed that idea because of my well-known views advocating exaggeration. It was no doubt courageous for him to be so blunt to my face, which gave me an opportunity to tell the rest of the story—backed up with the documents. He reserved his judgment for a few weeks, then sent me an e-mail apologizing for his harsh remarks at lunch. "I had never understood how your views were distorted—in fact I agree with them seeing your documents. I can only imagine how many others have thought as I did and didn't have the opportunity to find out the whole story from you."

I presume for every individual like him there are a hundred more who just take the false representations of my character and beliefs as true. I was told by a congressional staffer that because of this "double ethical bind" misquote, quite a number of congressional hearings chose not to ask me to be an expert witness—they feared I'd need all of my five minutes' time to clear the false charges and never get to my scientific substance. That such lies and innuendos could be effective in silencing my voice is perhaps the most painful cut of all.

CONTRARIANS IN THE MEDIA

A handful of contrarians, some of whom are neither scientists nor journalists, have challenged mainstream climatologists' conclusions. Remarkably, these public figures include the 2008 Republican vice presidential candidate, former Alaska Governor Sarah Palin. I say "remarkably" because presidential nominee John McCain was an early supporter of sound policies to prevent global warming. When I testified to his Senate committee on Science, Commerce, and Transportation in 2003 to support the McCain-Lieberman bill for climate policy, I was shocked that few members of his own party came to the hearing to support him—mostly Democratic senators showed up. It struck me that McCain, the maverick, was being punished by Republican colleagues for taking on an issue that their President and Vice-President denied, as did most members of his party in Congress.

A story in the *Washington Post* noted this irony of the global warming views of McCain's vice presidential selection: "A changing environment will affect Alaska more than any other state, because of our location," she said. "I'm not one though who would attribute it to being man-made."

Palin's comments stand in sharp contrast to those of McCain, who says at every campaign stop that he believes human activity is driving global warming.[11]

This is only one example of the expediency that drives those defending the status quo. When I'm asked about my opinion of the paradigm-altering claims of most contrarians ("After all, wasn't Galileo also dismissed by the establishment?"), I reply that we must carefully examine all claims, but only a miniscule fraction of such claims will overthrow well-established conventional wisdom like Galileo did. Nevertheless, these few contrarians are given disproportionate representation in the media despite dozens of official scientific assessments over three decades debunking their claims.

Receiving considerable recognition for conjuring up new and provocative contrarian "evidence" is Bjørn Lomborg, best known for *The Skeptical Environmentalist*, published in 2001. In a nutshell, Lomborg contends that the claims made by many natural scientists that large-scale degradation of the environment is taking place are false, or at least exaggerated; he says the "litany" being publicized by such people is not grounded in fact or solid scientific research yet it still infiltrates into the media. Lomborg, like Julian Simon before him, believes that the state of the environment is actually *improving* in most cases. Some specific examples he gives are: (1) acid rain has hurt lakes but barely impacted forests; (2) we are experiencing an increase, not a decline, in the services that natural ecosystems provide for us for free (for example, pollination, flood protection, food and fiber, etc.); (3) the 1989 *Exxon Valdez* oil spill was not nearly as bad as environmentalists make it out to be; and (4) biodiversity loss has been grossly exaggerated.

By far the most controversial chapter of Lomborg's book is the one on global warming. He states that money used to reduce global warming now would be better spent on reducing the burden of the poor by building hospitals, schools, and clean water infrastructure. He uses the Kyoto Protocol as a specific example, saying that the $80 to $350 billion he asserts it would cost per annum would only delay warming by six years, and would be better spent on dealing with the immediate problems of the world's poor. That same money, he says, could give clean water and better sanitation to the entire developing world, saving 2 million lives and preventing disease in another 500 million.

Lomborg weighs the costs of climate policy against the legitimate costs of sustainable development, as if they were the only two options—either cut greenhouse gases or leave the poor in developing countries to drink unsafe water. Is there only one either/or trade-off here? Why not make the trade-off between aid and the coal industry's free ride in polluting our waters and our air, or the trillion dollars in expenditures on the Iraq war? I, too, believe we should be investing in clean development—and the reality that the rich countries of the world have for decades underfunded this moral mandate is not a proud legacy. I argued that in *The Genesis Strategy* in 1976. But for Lomborg to lay at the doorstep of climate-protection advocates the label of devastating the world's poor is a new level of absurdity.

Perhaps what is most dismaying to me in all of this is that some media, like the *Economist,* and some governments quote Lomborg as justification for not embracing climate policies—when in fact they don't increase their investments in sustainable development either. Lomborg has become a ready justification for the forces of benefiting from the status quo to keep on denying the need for serious climate policy, this time by claiming they are helping the poor.

Many other scientists agree that Lomborg's work is deeply flawed. In fact, the Danish Committee on Scientific Dishonesty (DCSD) stated that his book was full of inaccuracies, ruling on January 7, 2003, that it fell within the concept of "scientific dishonesty." Opponents of Lomborg praised the committee's decision, while supporters were outraged. However, in what was a boost to Lomborg supporters, at the end of 2003, Denmark's science ministry in their new Conservative government rejected the DCSD's finding. As reported in a January 2004 *Science* magazine article on the subject, the ministry claimed that DCSD's findings [were] "flawed on several counts." The article went on to say that

[the ministry] held that DCSD's legal mandate is to rule on allegations of fraud, not on accusations of failure to follow "good scientific practise." It also criticized DCSD's ruling for lacking documentation, for failing to document the argument that the book is dishonest, and for describing Lomborg's research in unduly emotional terms. The ministry did not itself evaluate the soundness of the science or the claims in the book.[12]

How the DCSD could be accused of "lacking documentation" is inexplicable, given the voluminous testimony it received on errors in Lomborg's books. But the government that exonerated him had also just appointed him to a major environmental post, so to accept the ruling would have been an embarrassment.

One well-known Lomborg supporter was science fiction writer Michael Crichton. Invited to Caltech in January 2003 to give the Michelin Lecture, he presented a speech entitled "Aliens Cause Global Warming." In it, Crichton criticized mainstream scientists for attacking Lomborg's publisher, Cambridge University Press, in what he called the "new McCarthyism." He also deplored *Scientific American* for the article I co-authored entitled "Misleading Math About the Earth: Science Defends Itself Against the Skeptical Environmentalist," saying that the magazine attacked Lomborg for 11 pages and came up with few factual errors but only gave Lomborg a page and a half to write a rebuttal, which he contended was not enough space. He accused *Scientific American* of playing "Mother Church" and prosecuting poor Lomborg in the same way that Galileo was prosecuted for his novel ideas. (Likening Lomborg to Galileo is a very generous comparison.) Crichton was a gifted science fiction writer, but unfortunately, when he dipped into real science like climate change, his declarations turned out—like *Jurassic Park*—to be fictitious.

In an amazing turn of events, Lomborg recently dropped his nonsensical claim that climate change isn't a problem. Now, as if he thought of it first, he supports making "low-carbon alternatives like solar and wind energy competitive with old carbon sources," as he wrote in an op-ed piece in the *New York Times* on April 25, 2009. I guess scientists really are making progress in convincing people about climate change.[13]

LYING WITHOUT PENALTY

Media mud wrestling continues in the climate change arena. Among the most recent examples are the machinations involved in the production and distribution of the United Kingdom film *The Great Global Warming Swindle*. Purportedly a balanced documentary that is the antidote to the "distortions" of the IPCC, the film was shown in the United Kingdom by the sensationalist Channel 4 and across Europe—and reportedly had a major effect in weakening public confidence in global warming science. Among its claims is the absurd assertion that since carbon dioxide is only a tiny fraction of the atmosphere it can hardly be expected to have much effect. This was refuted by Australian climatologist Andy Pittman, who noted that an even more minuscule injection of Ebola virus would kill us, and that it is *effect*, not amount, of a substance that matters.

Another claim of the movie is that warming up until "now" wasn't unusual in the past 1,000 years—but what was labeled as "now" was a 20-year-old preliminary graph that did not include the radical warming of the past two decades, which, as noted earlier in the discussion of the "hockey stick," very likely exceeds all known warming over the past 500 years and likely over the past 1,300 years. To call the end of the graph "now" when it was really the 1980s is, frankly, a scientific lie. Similarly, the film claimed the sun could explain all warming and showed a very highly correlated set of graphs from 1500 to "now" linking global temperatures with sunspot cycles. What the movie's producers forgot to say was that the graphs left off the past two decades in which solar effects suggested cooling and the planetary warming went to unprecedented record levels—refuting their own theory. Even worse, the producers filled in a section of the graph to show a strong correlation several hundred years ago when in fact there was no data on it—they just made it up to look compelling.

The Public Broadcasting System in the United States refused to air the film, although it was shown in Australia at the insistence of the Conservative Howard government then in power—though handily trashed by an independent program that followed the broadcast, revealing its egregious distortions. Tony Jones, an iconic Australian reporter who anchors *Lateline* nightly, flew to the United Kingdom, interviewed the filmmaker Martin Durkin, and masterfully took him and the film apart step by step. It was one of the most adept pieces of science journalism I have seen, done by a political reporter who did his homework under the guidance of award-winning science producer, Annamaria Talas.[14]

A group of respected scientists and advocates filed a grievance against the film company and Channel 4 in the United Kingdom. They cited more than a hundred outright errors in the film and the deliberate misleading of some scientists who participated in it, and they claimed that its distorted presentation caused harm and injury to those who viewed it without access to the correct facts. Their suit was perfunctorily acknowledged—and the film company chided for several distortions—but in the end the complaint was denied in July 2008.

Banning films or filmmakers is flirting with censorship. But a succession of lies and misrepresentations without any rebuttal, on technically complex issues that most viewers would have trouble understanding even if there were a rebuttal, is a serious worry. I saw this film while sitting with Annamaria Talas in the Australian Broadcasting Company's offices in Sydney, and she cleverly filmed my reactions live on camera. Some of the distortions—such as calling 20 years ago "now"—were so egregious that I decried them as a "lie" in my shock. My on-camera spontaneous reaction was shown in the rebuttal that followed the movie's broadcast in Australia.[15] Because of the simultaneous rebuttal, no significant change in Australian public opinion on the seriousness of global warming occurred, unlike what happened in Europe.

Another who gets free license to distort is George Will, conservative columnist for the *Washington Post*. In his column published on February 15, 2009, called "Dark Green Doomsayers," Will wrote, "As global levels of sea ice declined last year, many experts said this was evidence of man-made global warming. Since September, however, the increase in sea ice has been the fastest change, either up or down, since 1979, when satellite recordkeeping began. According to the University of Illinois' Arctic Climate Research Center, global sea ice levels now equal those of 1979."

George Will in this statement lumped together many bits that do not support his conclusion that climate change hasn't affected the Poles. Indeed, the Antarctic sea ice has not changed much in extent, and the theory says it would do that more slowly than the Arctic sea ice. But recent analysis shows that even the poorly monitored Antarctic area is demonstrably warming on average, and fastest in the world in parts of the ice shelf regions of the Antarctic peninsula where several meters of potential sea level rise are stored as ice on the land. While it is true that the February area of sea ice in the Arctic is not very different over many decades, Will forgot to say that *summer* ice is at record lows and that both the Northwest and the Northeast Passages were simultaneously open for the first time the previous summer. In addition, the *thickness* of winter ice has declined dramatically over the past few decades. According to researchers with NASA and the National Snow and Ice Data Center in Colorado, more than 90 percent of the sea ice as spring 2009 began was only one or two years old, making it thinner and more vulnerable than at any time in the past three decades at least—since reliable measurements have been available.

George Monbiot, a British environmental writer and columnist for the *Guardian* who posts a blog at www.monbiot.com, was running a tongue-in-cheek contest to award the Christopher Booker Prize for the best media prevaricator. As he said, "The award will go to whoever manages, in the course of 2009, to cram as many misrepresentations, distortions and falsehoods into a single article, statement, lecture, film or interview about climate change." The award is named in honor of the *Sunday Telegraph* columnist Christopher Booker's "amazing ability to include misinformation and falsehoods in his pieces on climate change and the environment." Monbiot has kept an eye on George Will as a likely top contender.[16]

At the end of 2007 my old contrarian friend S. Fred Singer released a new book called *Unstoppable Global Warming: Every 1,500 Years.* Despite the topic of the book, he devotes only one page to the issue of what happens when carbon dioxide levels rise higher and higher. He quotes an authoritative-sounding paragraph from a "North Dakota climatologist" to the effect that rising CO_2 levels matter only at low concentrations and are of no consequence as the concentration gets higher. It turns out that the "climatologist" was actually a career lignite geologist in North Dakota. The paragraph was excerpted from his essay published in a four-page newsletter that circulates regularly among geologists in that area. The full essay explains that the geologist's source was a friend in the United Kingdom—who turns out upon further investigation to be a petroleum geologist. This is a telling example of the desperate lengths to which such authors must go in order to fight the overwhelming tide of scientific findings—hoping as usual that John Q. and Joan Q. Public will not look it up.

RESPONSIBLE REPORTING AND THE JOURNALIST-SCIENTIST-CITIZEN TRIANGLE

The media wars over climate change continue, although I think they are becoming less polarized than a decade ago. Perhaps Gore's *Inconvenient Truth* tipped the media balance. I hope we get some real scientific studies to analyze recent trends in media coverage of climate soon.

Yet the problem of falsely "balanced" journalism continues: polarizing an issue (despite it being multifaceted) and making each "side" equally plausible, mainly for the sake of simplicity but sometimes also to "sex up" a story by introducing bipolar conflict. Journalists need to replace the knee-jerk model of journalistic balance with a more accurate and fairer doctrine of perspective that communicates not only the range of opinion, but also the relative credibility of each opinion within the scientific community. Just as a good scientist records and analyzes all relevant data before reaching (often tentative) conclusions, a good reporter will delve deeply into an issue to ensure accuracy. Fortunately, most sophisticated science and environment reporters abandoned the model of polarization of two "sides," but this type of reporting still exists, especially since, in budget-cutting moves, mainstream media are firing their science specialists.

At one of Bud Ward's scientist-journalist exchanges, I recall a vice president from a big television news network defending "on business grounds" cutting specialists in science and environment. When challenged by the science journalists present, he retorted, "Any good journalist is trained to get the essence of a story in a day, and there is no problem sending general assignment reporters to cover environmental issues."

They argued for a bit before I jumped in. "So you think general assignment reporters largely trained in the doctrine of media balance can cover a complex topic like climate change in a day?"

"Well enough."

"So then I presume you send your general assignment reporters out to cover the Super Bowl," I challenged.

"That's different!" he snapped.

"Why, because it takes more specialized knowledge to distinguish a spread formation with empty back field from a Wildcat formation than to know whether the arcane cloud feedback processes that determine how sensitive the climate is to greenhouse gas pollution is relatively easier to explain?"

"It's just a business model that is driven by our stockholders and board, and we have to live with it."

"So do we," I snapped back. "So do we."

For their part, scientists can help by taking a more proactive responsibility for the public debate. They should help journalists by agreeing to weigh in on climate change debate. Also, scientists do not make their reputations by repeating what is well established, but rather by arguing at the cutting edge, where much speculation remains. Thus, any journalist looking for controversy has to go no further than one of our scientific conferences to find all they can scratch down on their notepads. What is often missing at our meetings are survey talks that summarize what is shared knowledge and well-established conclusions in the broad scientific community. We need to self-consciously start every session with a review of what is the state of knowledge so that those looking for controversy can put the cutting-edge debates into the context of the preponderance of existing evidence. It is not just the journalists who have work to do to improve the state of mediarology.

We could improve public dissemination of scientific knowledge if we required our science graduate students to take a survey course of the public communication process, including the process of political advocacy and science policy formulation. Similarly, journalism schools could show the consequences of misapplying "balanced" reporting techniques used in political arenas to complex issues in which not all opinions deserve equal billing in a story. A perspectives approach that elaborates on the relative credibility of many views on complex issues—not just the extreme opposites—is what is needed to properly inform the public. Literate citizens must take responsibility for educating themselves about all sides of the climate change debate so that they can see past biased opinions or bipolar "dueling scientists" models of reporting.

So how do we deal with this bubbling cauldron of special interests, paradigmatic misunderstandings, and time-honored and entrenched professional practices? While I don't have any simple answers, I do offer some guidelines that work for me—sometimes. First and foremost, we scientists must drop any superiority judgments; they only stiffen the resolve of those journalists who have been trained in their own profession's paradigms. Next, we should thoroughly explain how we arrive at our conclusions to those asking us for expert opinion. This explanation should include an explicit accounting of our personal value judgments, if we offer any. I do not hesitate to give such personal judgments when asked, as I, too, am a citizen entitled to preferences, but I always preface any such offerings by saying that my personal judgment is an opinion about how to take risks—not an expert assessment of the probabilities and consequences of future events. Third, it is essential that scientists go into explicit detail on their websites and in other places where depth is possible on how they arrived at their risk estimates. How did objective data contribute? How good was the data? What is subjective in the risk judgment? How did you arrive at the assessment?

These guidelines are not just for scientists—nonscientists should learn to ask scientists very similar questions: What can happen? What are the odds? How do you know? (These are the same questions I suggested to Senator Bradley in 1988 that he rightly stuck back in our faces in Q & A.) And where does your scientific assessment of risk end and your personal values come in? And citizens should learn to ask these questions not only on global warming issues, but on many other important socio-technical problems, such as medical protocols for dread diseases, in which risk methods and an explicit separation of facts and values are essential for best treatment. If Terry and I hadn't done that with my doctors to get best treatment for my mantle cell lymphoma in 2001, I very likely wouldn't have been here to finish this book or fight on for climate protection.

In addition, and now primarily for scientists, I often try to summarize what my colleagues say and publish, keeping in mind that scientific articles that have been through multiple rounds of peer review are far different from op-ed pieces, and are far different from individuals' congressional testimonies, let alone the say-anything world of the blogosphere. Perhaps most important, as Andy Revkin picked up in the *New York Times,* I encourage scientists to explicitly state what confidence levels they assign to their risk assessments and the degree of subjectivity needed to make that confidence label. How much weight a policymaker faced with limited resources might put on solving one problem versus another depends on many criteria, not least of which is how likely it is that the events might happen. Citizens should ask as well of experts: "How confident are you in your conclusions?"

Finally, I try to use accessible language and metaphors. Scientific jargon is effective for communicating with other scientists, but is often misunderstood in the public arena and increases the probability that a scientist will be "boxed in," misquoted, or, more likely, ignored altogether. For me, metaphors that convey both urgency and uncertainty are best—particularly for controversial cases like the more irreversible impacts of climate change.

For me the most important thing is to lay bare the issues I have learned in the trenches of the climate wars for all to see, so our democracy can be based on the most credible information. Our best safeguard, I have argued often, is to transition from "balance" to "perspective," where all significant views are reported and understood, along with their relative credibility. In such circumstances, we can place the necessary decision-making to deal with legitimate risks on a firmer scientific basis, built on the preponderance of evidence, not on a fictitious standard of total consensus of all who call themselves scientists.

One can always find a few scientists who will endorse an improbable apocalyptic vision such as extinction of humanity from climate change (to which I assign a virtual zero probability), or others who see CO_2 as plant food and warming as the elixir of a good life. Unfortunately there are Ph.D.'s with those completely marginalized views, and it is "buyer beware" on what one accepts as "certified" by experts—that is why each IPCC Working Group has 200 scientists, 1,000 reviewers, dozens of review editors, and three rounds of reviews. If that is too inconvenient for some bloggers, pundits, filmmakers, and special interests, then indeed there will be disconnects between them and most scientists.

This point was driven home to me in May 2009 at a Copenhagen World Business Summit on Climate Change, in which world business leaders got together with the likes of Al Gore, UN Secretary-General Ban Ki-moon, and Executive Secretary of UNFCCC Yvo de Boer, who presides over the Conference of the Parties. The summit was organized by the Copenhagen Climate Council, founded by its forward-thinking director Erik Rasmussen, a communications expert. Erik asked me to join in a panel on communications. He wanted me to present the ideas in this book to help the business delegates and international leaders understand why we've known pretty well for at least three decades of the climate threat, yet we were paralyzed from much action.

Erik also asked me to work with a youth movement he established, knowing that the opinions of 20-year-olds matter, as they are the generation inheriting our legacy. I spent considerable time with them, two of whom I already knew from Stanford, and we went all over the various issues. "So what is our most important message?" one asked. "Is it to limit the climate emergency we face by mentioning specific emissions reduction targets?"

I replied as best I could:

I think it is okay for you to discuss the science, impacts, and policy issues, but in truth that is not really your job now—that is the IPCC's job, among others. You have maximum credibility in telling my generation how you feel about their legacy to your generation. I'd tell them—were I somehow able to be 20 again while knowing what I know now—that you know your elders love you and want to leave you in a better world than they inherited. But the older generations' traditional model of "what was best for us is what is best for you" may not apply. You could say to them, "You were brought up to believe that the older generation has an obligation to leave us a legacy of wealth and infrastructure. We don't altogether reject that, but we are willing to trade off some of that consumptive orientation to get a legacy of clean air, a full complement of the diversity of nature and culture, and not just material wealth on a damaged planet."

... And most important of all, learn how to separate what part of the discussion is over scientific disputes and what part is over worldviews. Armed with that kind of literacy about sustainable development and communications, there really is a good chance you will have had a hand in getting the kind of world you'd rather have from those who can only change course if you tell them what you believe and what you value. Youth can be a powerful force for change through your honesty ... Always know some of us will be there right with you as you go through a lifelong apprenticeship in planetary sustainability management.

Without these kids, I don't know how I'd find the energy to stay focused in this 40-year war. Their caring honesty—along with my wife, friends, and family—helped me endure the three weeks in a hospital clean room during my 2002 bone marrow transplant. I learned firsthand the power of youth to focus the older generation to stay on track to achieve a better legacy. We have been making progress in changing people's minds, but we have a long way to go, and time is not on our side. As we'll see in the next chapter, the consequences of delay are likely to be dire, not only for humanity but also for all of the creatures that make Earth their home.

HABITATS FOR HUMANITY AND OTHERS

8 ⋮⋮⋮ **A FURRY LITTLE RELATIVE** of the rabbit called the American
⋮⋮⋮ pika, with bright eyes and round ears, is falling victim to
⋮⋮⋮ global warming. The pika lives in the alpine rocky slopes
known as talus high in the western mountain ranges of the United
States and southwestern Canada, where hikers have often been entertained by its high-pitched calls. A shy mammal with thick fur to protect it from the high-altitude winter temperatures, the pika eats a diet
of wildflowers and herbs. In order to provide food for the winter, the
little guys gather and sun-dry grasses during the summer, which they
store in small piles among the rocks, like miniature haystacks. But
they are among the most vulnerable of animals to global warming.

Surveys conducted a hundred years ago showed the pikas lived
around 7,000 feet in altitude, but more recent surveys show their
range now is closer to 9,000 feet, coinciding with the degree or so of
local warming over the past century. Because they already live at or
near the top usable habitat of the mountains, the adaptive behavior of
moving farther upward to a cooler spot is often not possible. Travel-
ing across lower valleys to relocate on another talus slope is virtually
impossible for these small creatures. Their heavy coats make it even

more thermally stressful for them to move substantial distances when temperatures rise.

The World Wildlife Foundation (WWF) has taken an interest in the fate of these endangered "rock rabbits." According to recent research by U.S. Geological Survey ecologist Erik Beever, global warming appears to have contributed to local extinctions of pika populations during the last part of the 20th century. In the Great Basin area—the area between the Sierra Nevada and Rocky Mountains—American pikas disappeared from 7 of 25 studied areas. WWF is funding the resampling of these 25 pika areas to find out how they vary across shorter time periods. The pika might be likened to the proverbial canary in a coal mine.

The IPCC Fourth Assessment Report in 2007 included serious language about the threat to biodiversity. It explains the nature of their plight and that of many other comparably threatened species:

> There is high confidence that climate change will result in extinction of many species and reduction in the diversity of ecosystems. Vulnerability of ecosystems and species is partly a function of the expected rapid rate of climate change relative to the resilience of many such systems. However, multiple stressors are significant in this system, as vulnerability is also a function of human development, which has already substantially reduced the resilience of ecosystems and makes many ecosystems and species more vulnerable to climate change through blocked migration routes, fragmented habitats, reduced populations, introduction of alien species and stresses related to pollution.[1]

We tried to express the scientific facts based on hundreds of studies in crystal clear language. "There is very high confidence that regional temperature trends are already affecting species and ecosystems around the world and it is likely that at least part of the shifts in species observed to be exhibiting changes in the past several decades can be attributed to human-induced warming." Of course, climate is but one factor threatening species—land use, exotic invasive species competing with natives for habitat space, and introduced chemicals have been demonstrated as interacting pressures by many existing published studies. However, these stressors are not always competitive, but can be synergistic—reinforcing each other.

We also expressed high confidence that the extent and diversity of Arctic and tundra ecosystems are in decline and that pests and diseases have spread to higher latitudes and altitudes. The problem is not just that polar bears will be losing their hunting areas to melting sea ice. Invasive species of plants and animals, including insects, that previously could not have tolerated the colder temperatures will move poleward, stressing preexisting ecosystems of Arctic plants and animals.

Each additional degree of warming increases the disruption to the structure and functioning of ecosystems around the world. Individual species have different specific thresholds to temperature, precipitation, and other environmental variables, beyond which they are not adapted; individuals risk death and populations may be threatened with extinction if they cannot move.

As the magnitude of climate change increases, further warming will almost certainly cause additional adverse impacts contributing to biodiversity losses. For example, only about half a degree of additional warming can cause harm to coral reefs. A warming of 1 degree Celsius (1.8 degrees Fahrenheit) above 1990 levels would result in nearly all coral reefs being bleached and species being rearranged in 10 percent of global ecosystems. Even worse, a warming of 2 degrees Celsius (3.6 degrees Fahrenheit) above 1990 levels will result in mass mortality of coral reefs globally. Species in one-sixth of the planet's ecosystems will be shuffled like a deck of cards, and about one-quarter of known species committed to extinction. By "committed to extinction" the IPCC meant that a long-term continuation of disruptions to ecosystems' structure and functioning could create situations where it would be too late for mitigation or adaptation to save many of those threatened species.

We have to take action to prevent that rise in temperature that exceeds the adaptive capacities of many species and the attempts of conservationists to configure biological refuges and interconnections between them to facilitate climate-induced forced migration. An increase above 3 or 4 degrees Celsius (5.4 or 6.6 degrees Fahrenheit) over 1990 temperatures, according to the IPCC 2007 report's summary of the scientific literature, carries a more than 95 percent chance of enhancing extreme weather events such as fire, flooding, drought, and heat waves, all of which will increase the disruption of ecosystems and loss of species.

In the April 29, 2009, issue of *Nature* magazine, I published at their invitation an article entitled "The Worst-Case Scenario," in which I explored what a world with 1,000 parts per million of CO_2 in its atmosphere might look like in the year 2100. Warming would be much greater than 2 degrees Celsius (3.6 degrees Fahrenheit). Human life would be vulnerable in many ways. Cities on river deltas or close to coastlines, particularly in Asia, would suffer because of rising sea levels and intensifying tropical cyclones, potentially creating hundreds of millions of environmental refugees and consequent political instability. The extent to which environmental refugees would create a military security problem is controversial, but a group of retired high-ranking U.S. senior military officers when asked to examine the role of climate change as a security threat called it a "threat multiplier." They meant that climate doesn't by itself create stressed human conditions but that climate change can be the final straw that breaks the already stressed systems and thus multiplies the threats.

Valuable infrastructure systems in densely populated cities worldwide could be damaged or lost in widespread flooding. The elderly would be at risk from unprecedented heat waves. Food stresses are a particular concern in areas that are already prone to droughts and poverty-stricken.

Some people would be at much greater risk than others, including many indigenous peoples, poor people in hot countries with little adaptive capacity, and those exposed to hurricanes and typhoons, wildfires, or flooding. The elderly and children with asthma or other lung ailments would be particularly affected by urban air pollution or wildfire smoke plumes exacerbated by another degree or so of warming, let alone many degrees more than that, as in the "worst case."[2]

After the article appeared, a reader from Colombia named Felipe sent me an e-mail asking what forest landowners like him could do to make a positive change. Felipe said that he was looking into forest protection incentives that would provide some benefit for avoiding deforestation of his land (200 hectares, or 494 acres, of native tropical forest). He asked good, tough questions: "Does it make any economic sense to own and protect a tropical/native forest land? Why? Can one expect to make a real living out of it? I understand their environmental importance, but here in Colombia in economical terms it seems nonsense, which is a shame and a contradiction on our planet climate change."

My reply to Felipe and others is this: "Until the nations of the world agree on a mechanism for putting a price on carbon and getting it to people on the land like you, your incentives are perverse—to destroy the forest for immediate income. Of course then it has no further value except low-intensity agriculture, and thus the return is unsustainable for a century until the forest grows again—if that even happens."

So it is critical that we succeed in finding mechanisms to allow people like Felipe to realize a sustainable income by protecting forests rather than getting a fast buck by destroying them. It will take foresighted and cooperative international governance to do this—and I am hopeful but not at all certain this will be achieved any time soon. In other words, I replied in a second set of e-mails, "Felipe, hold out as long as you can to at least keep the sustainability option open to you, and watch the world negotiations as they evolve over the next few years."

Felipe and others are betrayed by the way we mismanage the world with short-term financial disincentives to sustainability—yet some call it "progress," "development," and "growth." It is a standard prejudicial framing to refer to unmodified land in the U.S. as "unimproved," as if our disturbances are an improvement on nature's designs. We have to come up with better solutions.

One possibility is the Reducing Emissions from Deforestation and Forest Degradation (REDD) concept, which suggests that both biodiversity protection and CO_2 management in forests can be linked issues. This requires a determination of what the UNFCCC calls "additionality," that is, an investment now to protect forests *in addition* to what would have happened without a climate initiative. If it is determined that a particular forest protection investment was additional, then those instituting such forest-protection plans would be paid the going price for carbon emissions saved by the act of protecting a forest that otherwise would have been destroyed or degraded. This is the problem of the "baseline"—that is, a counterfactual guess as to what the rate of deforestation would have been in the absence of climate policy, and actions that reduce deforestation to a rate below the baseline would be eligible for "tons of emissions avoided" payments. There could even be a bonus payment for biodiversity protection, but as far as I know, that has not yet entered into the debate. The key problem is producing that counterfactual baseline against which we measure success of the project—in essence, how can we predict a future rate of deforestation before the fact? One of the details that must be worked out at the various UN-sponsored Conferences of the Parties (COPs)—Kyoto was COP 3 in 1997 and Copenhagen COP 15 in 2009—is to determine how to pay those who, like Felipe, want to be paid for protection, not destruction, of such valuable natural assets as primary forests.

Some in the EU don't like REDD, fearing it will be cheaper to protect forests than to change energy systems, and the search for renewable energy replacements could be delayed. The construction of new coal plants might continue, since the tons of CO_2 they emit could be off-set relatively cheaply by deforestation protection—for a while at least. Moreover, as the IPCC has showed, the amount of carbon it is maxi-mally possible to store in forests is well below the potential amount we would release if we burned a substantial amount of known coal reserves, like the 1,000 ppm "worst case."

This suggests that forests can be part of a climate solution, but they cannot be expected to mitigate more than a few tens of percent of even-tual CO_2 emissions if anything like a doubling or more of CO_2 occurs. All of this means that mechanisms to establish prices need to take into account the entire range of issues. This can be accomplished by willing parties, but not likely in only one or two successive COPs; it will prob-ably take many years to establish the rules most can live with.

I fought hard for such a framing at the Conference of the Parties 6 in The Hague in 2000, but was opposed not by the usual suspects—industrial interests and OPEC—but rather by those who were more "green"—World Wildlife Fund, Greenpeace, and European Green Party delegates. I was dumbfounded. Why didn't they want to support a plan to both keep carbon in the forests and get a double bonus of biodiversity protection? The debates were heated. I thought the argu-ment against it—no baseline for additionality—was legitimate, but not an insurmountable obstacle. Baselines are negotiable, and protect-ing primary forests should at least have been on the agenda. The pas-sion of the opponents seemed totally misplaced.

One evening during COP 6, I went to the environment NGOs' tent for a reception. In this more informal setting, I asked many of those attending what they were thinking. Finally, I understood. They wanted to punish the United States. "How so?" I asked.

"Because if we allow this relatively low cost mechanism, it will allow the U.S. to keep not cutting its emissions by mitigation, and anything that sanctions their refusal to take on deep emissions cuts endangers the world."

"But a ton of carbon is a ton of carbon," was my rejoinder, "and it doesn't matter if it is from retiring a coal-burning power plant or avoiding deforestation—and what about the double dividend of biodiversity protection?"

"We simply can't let the U.S. find any excuse not to cut its industrial emissions."

Okay, I got my answer—the biggest emissions scofflaw should do penance. But in the process of this ideological rant they were ignoring a critical strategy: to provide incentives to those in developing countries facing poverty to refrain from forest cutting. We needed mechanisms to get those places into the game, and a price on carbon for primary forest protection in return for a payment on the carbon saved was a way. The good news is that WWF and other such groups have new leaders, including Richard Moss at WWF—who is intimately familiar with all the climate policy perspectives from his years at IPCC. They have persuaded most environmental groups to incorporate carefully constructed primary forest protection into the COP process. We'll have to wait to see how well this can be done given the large amount of cash transfers that are determined by the baseline negotiations yet to be decided.

THE HUMAN COST OF CLIMATE CHANGE

Global warming increases the likelihood of drought and famine in drier areas of the world and flooding the coastlines of others. Certain indigenous peoples are in danger of losing their homelands. In the past few years I have been meeting with representatives of small island states, such as the Maldives, Samoa, Tuvalu, Kirabati, and Tonga, and high mountain areas in the Peruvian Andes. Although their ecosystems are vastly different, their problems are similar. Their leaders are deeply concerned about the future of their homelands, a future that seems no longer under their own control. International policies and pollutants are taking away their sovereignty and endangering their social, cultural, and economic welfare.

In Buenos Aires in 1998, a tribal elder from one of the small island states was offered a solution to their problem (losing everything due to sea level rise) by a diplomat of an OPEC nation. The island natives in his country would be paid to relocate to another part of the world, schools and hospitals would be built, and "the people of Kiribati would be better off." "But what can replace the submerged bones of my ancestors?" the elder chief asked the diplomat.

"But you are only a trivial fraction of world GDP," the OPEC representative replied, "and your situation is too tiny to be allowed to prevent the billions of poor in the world elsewhere from developing with the cheapest available fuels—coal and oil." So once again it is a "one dollar, one vote" cost-benefit argument used against a deep value judgment about fairness—when one group does most of the polluting and receives less of the consequential harm, and another group that did virtually none of the polluting pays the major price—not primarily in dollars they don't have, but in cultural survival. These are primarily ethical, not economic, issues, and they will be front and center in the global negotiations.

"But you are wrong and the OPEC diplomat was right," an economist in the audience at a talk I gave recently insisted. "Economics is a rational allocation system based on utilitarian principles—you know, the greatest good for the greatest number."

"Indeed I do know the utilitarian principle, but do you recall who was the first utilitarian?" He didn't. "Jeremy Bentham," I replied, "and what did he mean by utilitarianism: greatest good for the greatest number of what? Please fill in the blank."

A student in the first row did just that. "People," she rightly said.

"Yes, but in the discounted and dollarized cost-benefit analysis that you call rationalism, the market form of utilitarianism is the greatest good for the greatest number of what? Fill in that blank, please."

"Dollars," said the sharp kid.

"I don't see a moral equivalent between the greatest good for the greatest number of people and the greatest good for the greatest number of dollars," I insisted. "Is it rational to give a billionaire 10,000 utility votes for one vote of the average citizen? I repeat, this set of issues is primarily ethical, not economic." I don't think the economist's subsequent silence was acquiescence, but that is all part of the process of airing a broad range of views.

The small island states account for only a percent or so of the gross national product of the world. In economic terms, their loss would barely be noticed. But in terms of the injustice of sacrificing centuries-old cultures and vibrant societies to the self-interests of major greenhouse gas emitters, it is a human tragedy.

Take, for example, the Maldives, an archipelago of 1,190 islands in the Indian Ocean. The average elevation of the islands is only 1.2 meters (4 feet) above sea level. If global sea levels rise even slightly over the course of this century, which scientists predict as likely, most of the Maldives could be submerged. The problem is further exacerbated by the destruction of coral reefs, which formed the individual islands, by the acidification of warming seawater—another destructive synergism (warming and acidification multiplying their harms). Coral growth, in addition, may not be able to keep up with rising sea levels. And what happens to the 350,000 people who live on those islands?

The president of the Maldives, Mohamed Nasheed, was elected in November 2008 with a campaign promise to look for a new home for his country, should the expected rise occur. He is now scouring the world to find a place to move his 350,000 people, initially looking at India and Australia. He has been taken seriously, and Al Gore used Nasheed's initiative in testimony before the U.S. Senate Foreign Relations Committee as an example of what could happen if they failed to pass legislation reducing carbon emissions. It's a huge challenge to move hundreds of thousands of people from their ancestral home to a new area purchased or otherwise obtained from another country where the culture, economy, and environment are very different. A mass exodus is not supported by all the factions in the Maldives—some feel that Nasheed is crying catastrophe and embarrassing the country.

In March 2009 the president announced that he planned to make the Maldives into the world's first carbon-neutral country. He didn't specify details of how this would be implemented, but it set an optimistic tone. Perhaps his efforts, by example, would push Western countries into more aggressive policies of reducing carbon emissions, he said.[3]

At climate talks held in Bonn, Germany, in April 2009, a number of developing countries, including China, Bolivia, and the Philippines, insisted that developed countries cut their emissions rapidly and by far more than they had planned. The lead negotiator from the Maldives proposed that if the United States talks about ambitious targets, they would like to see a reduction of at least 45 percent below 1990 levels by 2020. That's quite a different goal from the present U.S. target of, at best, returning to 1990 levels by 2020—and probably less than that, as the 2009 congressional debate required compromises to get Democrats from fossil fuel states on board, and talks were mentioning targets as little as 6 percent cut below 2010 emissions. Most Republicans simply opposed such carbon limits.

But on June 26, 2009, a remarkable breakthrough occurred in the U.S. House of Representatives after intensive lobbying by the Obama White House and Democratic congressional leadership. By a slim margin of 219 to 212, it passed the contentious Waxman-Markey climate and energy bill—despite many weakening amendments to soften the blow for many industries and agricultural interests. Nevertheless, this historic first set out aspirational targets for emissions reductions that start slowly (17 percent cut below 2005 levels in 2020), but accelerate rapidly over time (83 percent cut in 2050).

Despite the amazing rancor over the targets, to me the numbers that really count for the next ten years are not so much the 2020 targets, but the financial commitment that we will devote to the problem. How many tens of billions of dollars a year are we firmly committed to spend on both new deployments of low-emitting technologies, carbon recapture and storage, and transfers of cleaner technologies in partnerships with developing countries? Sure, aspirational targets are necessary to scale roughly how much to invest, so I am not against the targets per se. But without explicit policies and measures to achieve any aspirational targets, it is talk, not action. Remember the problem with the Kyoto Protocol: targets without teeth.

Fortunately, the Waxman-Markey House version of the bill had many explicit measures, such as efficiency standards and investments in clean technology, as part of the package. This complexity led to a bill of more than a thouand pages, making the process of figuring out exactly what was in and out at any one moment in the bargaining process very difficult for those following the acrimonious debate in real time.

Meanwhile, the poorer countries are both serious and united. The Bolivian ambassador described the issue in no uncertain terms. "Developed countries have over-consumed their share of the atmospheric space—they ate the pizza and left us the crumbs."[4]

I don't blame the Bolivians for being mad. Their Chacaltaya glacier has disappeared. It began melting in the mid-1980s and by March 2009 was completely gone. The glacier, whose name means "cold road," was 17,388 feet above sea level, and it was 18,000 years old. The team of scientists who studied the glacier from the Institute of Hydraulics and Hydrology at the Universidad Mayor de San Andres in La Paz, headed by Edson Ramirez, had concluded ten years ago that the glacier would survive until 2015. But the rate of thaw was triple what they expected. Anthropogenic global warming appears to be the primary cause. Other glaciers in Bolivia, Peru, and Ecuador will suffer the same fate. Not only will it affect the tourism industry (no ski resorts now), but it will also endanger the water supply for the millions of people who live on the western, mostly arid side of the Andes. Rain, snow runoff, and melting glaciers are rapidly reducing their water supply, and the melting process can trigger the opposite concern—the meltwater backs up, held by an ice dam that eventually bursts, causing serious flash flooding downstream. So these inhabitants face both floods in the short run and drought later on—and what fraction of global emissions did those so affected produce themselves?

Equally as vulnerable are the indigenous peoples of the world. They depend upon the natural environment for their homes, livelihood, and traditional cultures. They are on the front lines of climate change, and they were among the first groups to call upon international organizations such as the UNFCCC to protect the planet from human-caused climate change.

In April 2009, the Inuit Circumpolar Council hosted the Indigenous Peoples' Global Summit on Climate Change in Anchorage, Alaska. Delegates from all regions of the world came together to share their knowledge and experience in adapting to climate change. The main focus was to develop key messages to deliver to the world at the Conference of the Parties in Copenhagen in December 2009. Miguel D'Escoto, president of the UN's General Assembly, received a standing ovation when he signed the Declaration on the Rights of Indigenous Peoples and the Action Plan developed at the summit. The attendees hope it will be used as a road map on how to tackle climate change and who will be involved.

A Navajo delegate emphasized that the world needs to get to the root cause of why global warming is happening, in order to develop equitable solutions. "People are totally unaware when they switch on their lights what it takes for their light to come on. Coal is mined where I come from, it's transported through someone's territory, it's burned in someone's territory, waste goes to someone's territory." Not only are indigenous peoples worried about oil drilling and coal mining in their homelands, but also about ice melt, diminishing food sources, sea level rise, drought, and threatened migration of communities forced by the climate change.

The Inuit Circumpolar Council represents 150,000 Inuit worldwide, and the conference they hosted was attended by a cross section of the estimated 5,000 distinct groups of indigenous peoples identified in more than 70 countries, with a combined global population estimated at 300 to 350 million. The unified voices of these people must be heard.[5]

In January 2009, Terry and I were honored to be invited by Aqqaluk Lynge, head of the Inuit Circumpolar Council of Greenland, to discuss with the Inuit leaders the strategies for adaptation to climate change and to enhance their voice at international negotiations like Copenhagen 2009. Greenland is on the front line of global warming and already is in the hot spot of glacial melting.

Aqqaluk met us in Illulissat, a town at the outflow of the fasting moving ice stream, known as the Jacobhaven ice stream. It has receded in a hundred years about 50 kilometers (31 miles) as ice formerly grounded on the floor of the fjord melted, and now the Greenland ice sheet calves almost directly into the fjord upstream of Illulissat, dumping massive icebergs into the ocean, raising the sea levels. We took a cruise into the fjord and were jostled by numerous small ice floes, reminding me of bumper cars at an amusement park. These beautiful fantastical shapes are direct evidence of ice melting, but other than the recession of the ice stream back toward the mainland, it is arguable how much faster this system is calving icebergs into the oceans than before global warming melted the grounded fjord ice stream. But NASA satellite surveys show the Greenland ice sheet is melting at its margins much faster than small snow buildups are occurring on the center of the sheet, and that melting is accelerating.

The Greenland ice sheet now calves almost directly into the fjord upstream of Illulissat, dumping massive icebergs into the ocean, raising the sea levels.

Terry and I both teach at Stanford, and we believe that a university relationship with the Inuit Circumpolar Council of Greenland is a highly desirable educational opportunity. All we're waiting for now is to find the resources to bring Greenland Inuit leaders to Stanford to meet the professors and students, then send a few of those academicians to Greenland over a summer. The goal is to gather information that will help the Inuits make their inevitable transition to a melting world. We traveled by small plane to the south to Nuuk, the capital city previously known as Godthåb. Terry has already set up a protocol with Aqqaluk's sustainability head, Lene Holm, for Inuit hunters to record data such as the size of hunting parties, the individuals' proximity to each other while hunting, and various measures of their prey that will serve as a census for polar wildlife. These hunters really know what they're seeing, and we want to capture and codify that knowledge.

We did an interview with a national newspaper journalist—a weekly paper written in both Danish and Greenlandic languages. I never thought I'd end up in Greenland television studios, but one remark I made that got into print set it off. When we landed in Greenland, we saw some men carrying what looked like ice coring equipment. So Terry asked one of them which ice research team he was on, presuming it was a scientific study group, which is not a rare sight on the ice sheet in summer. It turned out they were working for Alcoa Corporation and they were prospecting for possible mineral exploitation sites now that the ice sheet was receding. It was classic—opportunity for economic development, side by side with the threat to indigenous culture as the traditional hunters were struggling given the ice recession from warming. I told the reporter to just imagine what it would do to the culture of the roughly 70,000 Inuit living in Greenland when a big mining operation comes in with 10,000 rowdy workers from Western culture getting pie-eyed on Friday and Saturday nights.

I suggested that if they make a deal with these international mining companies like Alcoa, the companies be required to hire and train a certain percentage of Inuit to do the work. Also, I pointed out that mining often has serious air and water pollution issues, and thus the Greenlanders should make sure that the companies pay for independent third-party monitoring of air and waters to be sure it was safe, so they would not become an Arctic version of West Virginia's mountain-top removal environmental disasters. As usual, we told them about climate change and why Greenland was the front line, or as Terry likes to say, "ground zero." That was not really news to them—they had long ago accepted the reality of climate change. Yet fewer had considered the cultural and environmental side effects of inviting big companies to operate there.

The Greenland Parliament is now starting to consider this issue, including having a system for monitoring air and water pollution, and it will need to be carried out by an independent party. The elders also have expressed tremendous concern about what this boom would mean for the existence of the Inuit as a distinct culture. Although they are a tiny fraction of the world population, they are a big component of a unique culture. The Inuit will have a tremendous opportunity for mineral exploration, and how they handle it is going to determine the culture's evolution for a long time. Nevertheless, this climate trend looks like a disaster for the subsistence hunters.[6]

By the time we left Greenland, Terry and I were even more committed to ensuring that the indigenous peoples would be fairly treated in future decisions. No community should be forced from their home or their culture—whether a tropical reef island or a once frozen tundra. We must work together for both cost-effective and equitable solutions.

WHAT THE BIRD-WATCHERS KNOW

My wife, Terry Root, is currently investigating possible ecological consequences of rapid warming to birds and other species around the globe. As a world-renowned scientist who has received numerous awards, Terry is no one to mess with, especially by those who claim species' fate doesn't matter.

When she was doing her Ph.D. work at Princeton, long before we crossed paths, Terry already was an innovator in avian ecology. When most ecological studies were done in areas about the size of two tennis courts, her study area was the entire North American continent. Other ecologists found competition, predator-prey, or other biotic interactions the primary forces influencing species at the smaller scale. Her maverick insight—that factors affecting species were being missed at that small scale—led her to determine how important the broad-scale environment is for birds. Her way of thinking was seen as unusual because the dominant culture held that real ecologists had to collect their own data, with researchers' personally verifiable results from their own small research plot. A veritable army, however, is needed to collect data throughout North America.

The army Terry used was made up of bird-watchers taking part in the National Audubon Society's Christmas Bird Count, when people go to areas around the country and record the number of individuals of each bird species seen on a day near Christmas. Her analyses of these data showed that in the winter a vast majority of birds go as far north as they can before the nighttime temperature gets so cold that they cannot survive throughout the nights.

After showing this in North America, she incorporated data from many other scientists around the world into her analyses, in order to posit hypotheses about spring events, such as the blooming of tulips or nesting of robins, occurring in species around the globe. Her global analyses in many ways revolutionized her field.

Being a maverick guarantees that some do not approve of what you do, and this was true of Terry's work. One of her advisers at Princeton refused to sign her dissertation. "You must collect your own data to do real science!" he said. (Hand-to-hand combat is not just for climate scientists.) But she was strongly supported by her primary adviser, Robert May, who later became science adviser to two British prime ministers and head of the Royal Society, and John Weins, who had been following her dissertation work from afar; both praised the effort and signed the dissertation.

Terry was one of the earliest biologists to study the effect of global warming on plants, birds, and other animals and to determine the physiological mechanism by which long-term, large-scale change could happen. For her it was not a big leap to her next set of studies, but again this work was very unusual for biology. She used published studies of plant and animal movements from around the globe to see if they showed a consistent pattern worldwide, which you would expect if global warming were causing it.

The method she was using is called meta-analysis. Meta-analyses provide methods for combining results from various studies, whether statistically significant or not. Results from meta-analyses determine whether there is a consistent "signal" or "fingerprint" among the studies. The balance of evidence from her meta-analysis done on species from many different animal or plant groups examined on six different continents around the globe suggests that a significant impact from recent climatic warming is discernible in the form of long-term, large-scale alteration of animal and plant populations. In 2003 Camille Parmesan at the University of Texas and Gary Yohe at Wesleyan University published a similar meta-analysis to the one Terry's team used for the IPCC Third Assessment Report in 2000. Even though these studies had some significant methodological differences, their basic conclusions were very similar. Terry's study from 2000 for IPCC was extended, improved, and published in the same issue of *Nature* magazine as the Parmesan-Yohe meta-analysis—which enhanced the impact of both studies.

For Terry's team study published in 2003, information was gathered on species and global warming from 143 studies, each of which spanned at least ten years. The study showed a trait of at least one species that exhibited change over time, and it found either a temporal change in temperature at the study site or a strong association between the species trait and site-specific temperature. To document a strong role for climate change in explaining many of the observed changes in animal and plant populations, Terry and her colleagues looked for repeated examples occurring over long temporal and broad spatial scales. The predicted result, or fingerprint, of an underlying consistent shift in a large-scale pattern shown by many species around the globe, coupled with an understanding of the possible causal mechanisms, provides high confidence in attributing observed species changes to climatic change.

Their study shows that recent temperature change has already had a marked influence on many species, ranging from mollusks to mammals and grasses to trees. Of all the species showing change, more than 80 percent of the species shifted in the direction expected by global warming based on the species' known physiological constraints. For example, their analyses showed an average shift in spring timing of events, such as breeding or blooming, for temperate-zone species of about five days earlier in a decade and that species at higher latitudes are shifting poleward as the temperature climbs. The conclusion that significant impact from recent climatic warming is "discernible" was combined with work from lead author Cynthia Rosenzweig by the IPCC in our Third Assessment in 2001 to include "environmental systems," such as melting glaciers. Clearly, if such climatic signals are now being detected with a warming of "only" about 0.6 degree Celsius (1.1 degrees Fahrenheit) up to the year 2000, the expected impacts on environmental systems and ecosystems of temperature increases up to an order of magnitude larger by 2100 will likely be dramatic.[7] For example, Terry found that detailed records in the Upper Peninsula of Michigan showed that 30 years ago, the mourning dove, a fairly common species, use to return on spring migration in the middle of April, yet 26 years later it had expanded its range northward to become a year-round resident. Of all the species monitored there, about a third exhibited changes, and all but one of these changed in the expected direction—arriving earlier from migration as the climate warmed.

Like the furry pika's abandoned haystacks in the talus slopes, the movement of birds, plants, and animals northward away from the warming temperatures of their natural habitat are signs of the changes already in store for Earth. Our responsibilities as human stewards of these co-inhabitants are clear and compelling.

A VOICE MUZZLED

When we talk about biodiversity, one country stands out. A mountainous archipelago once dominated by temperate rain forests, New Zealand possesses a biodiversity that is unique to the world. Eighty million years ago the islands separated from the mainland, and after that event, the natural species developed in isolation—until the first contact with humans a thousand years ago. Before that, only two species of mammals were native to New Zealand, both of them bats.

Many of the other species of flora and fauna were distinct to the islands. Some birds, most notably the flightless kiwi and a ground parrot, evolved to take advantage of the fact there were no terrestrial mammals. When the ancestors of the Maori people first arrived, they brought with them rats and dogs, which became predators of the native wildlife. Captain James Cook arrived several hundred years later with more rats, dogs, pigs, and goats. In addition, some 15 percent of the flowering plants on New Zealand are not native. All of these introduced species and others such as the weasel-like stoat have had a major impact on the biodiversity of New Zealand, especially on endemic (native only to New Zealand) birds and plants, which are now being further imperiled by climate change.

The changes in habitat for native species have increased rapidly due to deforestation, draining of wetlands, grazing, hunting and fishing, and degrading of ecosystems by invasive species of plants and predators. The government of New Zealand implemented a formal Biodiversity Strategy coordinated by the Conservation Department. It has had many successes in restoring wetlands and protecting endangered species. The government elected in 2008 appears to be less conserving and more conservative in its views on global climate change.

Jim Salinger, easily the most recognizable climate scientist residing in New Zealand, whom I've known since 1978, worked for the government science agency, the National Institute of Water and Atmospheric Research (NIWA). For three decades he has been a media spokesperson and the public face of the organization. He's been a lead author of IPCC many times, heads a World Meteorological Organization program in agricultural meteorology, and frequently testifies to parliament. He also is sure that anthropogenic climate change has potentially dangerous effects to the small nation and the world. His New Zealand friends include members of parliament (MPs) and governmental ministers.

During the spring of 2009, the government appeared to be trying to scuttle their climate policies from the previous Labour government, which had developed an emissions trading scheme to put a price on carbon and thus mitigate emissions somewhat. I was the first witness before the select committee on emissions trading that is rethinking the emissions policy of the previous government. The questions were somewhat skeptical in nature, and toward the end of the hearing they even hit me with that old contrarian chestnut, the *Discover* article misquote from 1989 about "the right balance between being effective and being honest," again omitting my stated hope that it meant "doing both." (Maybe I'll have that engraved on my tombstone). After I showed the dishonesty of that misquote and, often, those who use it, I think they were sorry that they tried to misrepresent me.

The night before, Jim had made an odd comment about maybe we should turn down the invitation to go to dinner with two ministers who wanted to talk with us. "Why?" I asked, perplexed.

"Well, it's a long story . . . I have been told not to talk to MPs without permission."

"Whose permission?"

"My CEO's."

"What, aren't you living in a free country with rights to assemble as you choose?"

"That's what my lawyer said, too," Jim reported.

"Sounds like Bush has moved down under," I quipped. "They can't be serious?"

After making light of this absurdity, we went to dinner with the ministers. Naturally, in the course of a long conversation, the issue of the next day's hearing inevitably came up and their perspectives helped me to understand the situation better and thus be better prepared to testify effectively.

A month later, Jim e-mailed me to say that he was in big trouble with NIWA. The administration had instituted another new policy that NIWA employees were not permitted to speak publicly without gaining prior approval. Jim had been accused of "serious misconduct" when he participated in a New Zealand television news story about the shrinking of New Zealand's glaciers. He had been told that he could not talk to the media or members of the government without permission. Specifically, he was also warned that he could not communicate with MPs without violating NIWA policy.

Jim was fired on April 24, 2009, ostensibly because he ignored NIWA policy on his lawyer's advice and had dinner with some friends— who happened to be members of parliament. Of course, all will eventually come out as Jim has challenged the decision in a legal proceeding, and I would expect that his detractors will have other reasons for justifying their decision that we haven't heard about yet. But regardless of the merits of Salinger's specific case, the suppression of a scientist's access to a media that is trying to do its job by finding the most credible spokespersons on particular issues conjures up deep worries about censorship. Attempting to censor their most famous resident climate scientist—a person I have always found to be of integrity and principle—smacks a bit too close to what happened in the Bush Administration. Here, statements from government scientists who disagreed with the administration's beliefs were changed by political appointees, and not by scientists.[8] As Jim said to the media, "It's not as though I'm doing bad science, it's not as though I'm not performing, and so I'm really astounded."[9]

The New Zealand media covered the story in sympathy with Jim Salinger's position, and he was lionized in editorial cartoons that made fun of his boss. When the media found out that access to one of their favorite contacts was being denied to them, they took umbrage and made it a national political disgrace in editorial after editorial. When a New Zealand reporter called me for a reaction, I said, "Managers are a dime a dozen, world-class scientists very rare. Maybe the wrong guy at NIWA got sacked."

NIWA says it will not talk about an employment issue publicly. But when climate scientists of Jim's stature and integrity are silenced by their own democratic governments, what can the rest of the world do?

My answer is, plenty. The management has been bombarded with angry e-mails from scientists around the world and stories in the international science press—all very embarrassing to the agency. On a grander scale, we can keep up the pressure for cooperation and justice at international climate conferences. Elect courageous members of government who will engage in the climate battles. Educate the public about the possibilities for mitigation and adaptation to prevent the worst-case outcomes of climate change. Take action while we still have time to be effective. Think about your children and grandchildren living in the world of the future, which is presently in our charge. That also is what Jim Salinger would answer, and that, in a nutshell, is what we all need to do.

WHAT SHOULD KEEP US AWAKE AT NIGHT

9 **DURING THE YEAR OF WRITING** this book, I have traveled tens of thousands of miles, from Greenland to Europe to Japan to New Zealand, and given dozens of lectures and interviews. I've participated in tribal councils, forums with religious leaders on planetary stewardship, summit meetings, academic conferences, scientific assessments, government testimonies, business roundtables, and private one-on-one meetings with industrialists and policymakers. We never did find a silver bullet to rescue us from the dangers of global climate change—although there is plenty of bronze buckshot, as climate policy analysts like to joke.

We continue to face two critical challenges. First, we must protect the planetary commons for our posterity and the conservation of nature. At the same time, we must also fashion solutions to deal fairly with those particularly hard hit by the impacts of climate change and national and international climate policies.

In front of a U.S. Senate committee hearing I once testified about "the question of the great uncertainties involved and the unfortunate overlap in the time scales for which present estimates suggest major effects and the time frame by which we may have certainty that CO_2

is a real problem. This will lead us directly to the question of its impact on society. The time it takes for society to adjust its strategy for something as important as energy production is at least on the order of decades and could well be on the order of half a century, judging from past experience."[1]

In a later testimony to the U.S. House of Representatives Ways and Means Committee I said: "What is called for, in my view, is another step in the sequencing of actions: public/private partnerships to foster the learning-by-doing projects to make renewable energy systems cheaper and more available and to explore other options from both cost and safety aspects. It is not just R & D, but . . . deployment of prototype systems to compete for future market share based on their improved performance gained from the demonstration investments."[2]

What is remarkable in these somewhat similar statements is not the common push for renewable alternative energy systems, or for security and environmental concerns to be aligned in a coordinated policy framework—this advice has been repeatedly offered for decades by scientists, policy analysts, Al Gore, Tim Wirth, and nonprofit groups. Rather, the first set of quotes is from a Senate hearing in 1979, and the second is a House hearing 28 years later in 2007. This three-decade period has witnessed too-similar pleas that have still not been translated into meaningful climate policy.

Almost a third of a century has passed since the U.S. Congress—and many other governmental authorities throughout the world—first learned of the climate change problem and possible solutions. If some of these solutions had been implemented back then, as Jimmy Carter argued in the late 1970s, we would now be further on the route to lower-cost clean energy alternatives than if we panic-buy them now. And if we wait even longer to act, we will have to do even more panic purchasing at a faster rate, driving up costs and lowering the feasibility of avoiding many dangerous climate impacts.

But worrying about water that has already flowed over the dam won't help us much now—or will it? The idea here is not primarily recriminations for past misdeeds, but rather learning if there are lessons in this history of failure to act on how to do a better job now. Sometimes my students try to comfort me: "You can at least say 'I told you so'!" "Nah," I reply, "an 'I told you so' is really an 'I failed you so'—we just didn't get it done."

CAN DEMOCRACY SURVIVE COMPLEXITY?

We have seen what blocked implementation of strategies for clean energy. The political chicanery of ideologists and special interests has misframed the climate debate as bipolar—"end of the world" versus "good for you," the two lowest probability outcomes—and the media often carries it in that frame. The confusion that bipolar framing has engendered creates in the public at large a sense that "if the experts don't know the answers, how can I, a mere lay citizen, fathom this complex situation?" In addition, the recruitment of nonclimate scientists who are skeptical of anthropogenic climate change to serve as counterweights to mainstream climate scientists also is an old trick to spread doubt and confusion among those who don't look up the credentials of the claimants—and that, unfortunately, includes most of the public and too much of the media. The framing of the climate problem as "unproved," "lacking a consensus," and "too uncertain for preventive policy" has been advanced strategically by the defenders of the status quo. This is very similar to the tactics of the Tobacco Institute and its three-decade record of distortion that helped stall policy actions against the profitable tobacco industry, despite the horrendous health consequences and eventually billions of dollars in successful lawsuits against big tobacco.[3]

In both cases—climate policy and smoking regulations—documents leaked by whistle-blowers later emerged showing that this "it is unproven" framing was no accident, but a well-financed campaign to spread confusion and doubt. Helping to make these tactics successful was giving their spokespersons equal time in media stories as "the other side," as if it were a bipolar "guilty versus innocent" question.

Even when big oil pretended to invest their big profits in new exploration, the truth emerged during the massive spike in petroleum prices in 2007 and early 2008 that resulted in record profits for companies like ExxonMobil—$40 billion in 2007 and $45.2 billion in 2008. Did these oil giants reinvest the money in efficiency, or exploration for more oil resources, or development of renewable alternative energy sources? Yes, to the tune of a few billion—big-sounding numbers *until* you learn that the vast bulk of their windfall profits was spent on repurchasing company stock on the market, driving up its price so those who held many shares—like the ExxonMobil board of directors and senior management—dramatically increased their personal portfolio values.

Perhaps you are not too surprised by this kind of greed and short-term thinking. Neither am I. What I worry about is not so much this all-too-common "me first" behavior, but rather how many decent people are still taken in by it. What keeps me up at night is a disquieting thought: "Can democracy survive complexity?"

If the public understood the basics of the real risks to nature and to themselves, their posterity, and their world, they would be much more likely to send strong signals to their representatives to act in a precautionary way. They might even learn how to vote consistently with their values on ballot propositions to solve the problems that the legislature and executive branch avoid. But if daunting complexity, fueled by deliberate special interest distortion and knee-jerk media balance, is what we hear predominately, then democracy has a hard time dealing with slowly evolving, large-scale, complex problems such as climate change and policy, health care policy, security and military policy, or education policy, to name a few.

SENATOR FOR AN HOUR?

I am often asked by legislators what I would do if I were in their place—what I call the "Senator for an Hour" question. After thanking them for the honor, I remind them that what needs to be done is risk management—weighing the risks of inaction versus the implications of proactive policies. That is a value judgment, and I usually lay out my worldviews at the outset.

There are two factors that imbue a problem with special concern for me. First, if a problem is caused by activities that confer benefits to only a few, but the risks are doled out to the many, including those who had little benefit from the behaviors that created the risk, then I have a higher priority to fix such a problem than for ones less inequitable. Second, if a set of problems conjures up a sufficient chance of irreversibility, and there is no policy reversal possible once we have passed some critical threshold or tipping point, then that too intensifies my attention.

I often surprise the senators, who by now expect me to say first we should tax the hell out of the polluters. Yet the first thing I would try to do is start an honest and transparent dialogue, so media, government officials, corporate leaders, and just plain folks understand what is really at stake. What would it take to adapt to and mitigate the risks? Who can and should pay for it, now and over time? Despite the "greed gene" that nobody has isolated yet in *Homo sapiens,* I think most of us can do the right thing for the collective good when we understand the situation. As John F. Kennedy said in his famous Inaugural Address, "Ask not what your country can do for you, ask what you can do for your country." Why is it that line has been so enduring if we don't also have an "altruism gene" mixed in our DNA along with the greedy one? In this case, though, substitute the word "planet" for "country."

Sometimes I find myself delivering this message in unexpected quarters. How did this boomer-generation scientist end up in the summer of 2008 in the pages of *Billboard* magazine talking about my interactions with hip-hop artist Michael Franti, the rockers of Widespread Panic, and rapper Snoop Dogg?[4]

A year earlier I had given a talk in honor of my old friend Warren Washington at NCAR in Boulder. A young graduate student approached me, not to ask the usual questions about science or the policy debates but about my interest in running a "think tank" at a rock festival in Michigan the next summer.

"You've got to be kidding!" I retorted to Shannon McNeely, the student. "I wouldn't know even one of them. If they had Bob Dylan, that's a different story—but rockers and rappers, no way."

"Listen, we want you to run a bunch of panels that put the stars together with serious scientists, journalists, business leaders, and politicians, so we can get Gen Y-ers involved. You are one of the few scientists who can adjust your style to the audience, so why not do something different with a community that needs to hear this stuff?"

Hard to argue with a smart kid, so I said I'd consider it.

Over the next several months we put together half a dozen panels, bringing music stars on board along with guys like Robert Parkhurst (who runs the multimillion dollar "Climate Smart" program for the Pacific Gas and Electric Company), along with energy heroine Hunter Lovins, hip-hopper with a positive message Michael Franti, and many others. The Rothbury Music Festival was held on the Fourth of July 2008 and celebrated green themes, requiring all participants to separate their wastes into trash for the landfills, compostables, and recyclables. Volunteers stationed at each recycling center showed everyone from high schoolers to Snoop Dogg how to do it. Nobody was a bigger star than Mother Earth that weekend.

One interesting sidelight was the deference shown by the rock stars on our panels. Superstar lead vocalist John Bell from Widespread Panic, in front of a small crowd of 300 people, admitted how difficult it was for him to offer an opinion when surrounded by so many left-brain types like Hunter Lovins and media ace Bud Ward and me. So I said, "If I put myself in your place, stuck between all these fast-on-their-feet accomplished public speakers, maybe it would be like me taking my 12-string guitar up with you in front of the 25,000 screaming fans you entertain every night and playing solo—knowing I'd be making three mistakes a minute—very intimidating. I'd be petrified!" He laughed, and more easily joined in—and was very effective in motivating the participants, not as a mock scientist but as an opinion leader on how important it was to *him* to push for planetary sustainability.

If we are to make progress in dealing with the environmental protection and economic development dilemmas, then events like this need to take their place along with community organizing, business leadership forums, state actions, federal governmental incentives, and international accords. There is no time to lose and no one should be glibly dismissed because he or she is not a scientific expert. We are asking them to join in finding the solutions, not to define the scientific risks. We shouldn't underestimate the impact of young people committed to the health of planet Earth.

STEPS TO MAKING CLIMATE POLICY

I like to argue for a series of steps in a sequence of making climate policy. The idea of "cap and trade" has received strong attention in California, Europe, Australia, and now the U.S. Congress. That would mandate a cap on emissions so that concentrations of greenhouse gases in the atmosphere could be managed at a safer level. The trade part would arise if one firm, say, a progressive power plant company, finds a way to cap their required emissions below their quota, leaving a cushion. They could then sell that cushion to others not so progressive to help them meet their quotas more cheaply than if the less efficient firm cut their own emissions directly. In this way, the already efficient have an incentive to get more efficient in order to have more cushion to sell. At the same time the inefficient firms have an incentive to get more efficient so they don't have to keep buying permits to pollute.

Environmental groups like cap and trade since it provides some indication of the levels of pollution permitted. Businesses, on the other hand, often are leery, since they cannot predict in advance how much it will cost. Such groups prefer a "carbon tax" on emissions, so they will know just what costs they face for a variety of their activities that emit carbon. But environmentalists counter that price certainty is not a guarantee of quantity certainty, which they want for environmental protection. Certainty over price versus quantity emitted is a classic trade-off between economic and environmental values.

However, these two schemes are not as different as the debate might seem at first, since a cap and trade requires investments in new cleaner processes—which cost real money. That gives rise to an implicit price on carbon known as the "shadow price." The key to this kind of market solution is to get a price on carbon to stimulate investment in less polluting alternatives and to alter behavior with regard to dumping wastes in the atmosphere for free. With a dumping fee—the carbon price—companies would finally have incentives to become cleaner as a routine component of doing business.

Such price increases would be partially passed on to consumers, but that means consumers can choose, based on price, more efficient cars, appliances, and other products, thus increasing demand for cleaner products. This scheme has worked in the United States very well to reduce emissions causing acid rain, so most economists like such "market-based" solutions. The political downside for politicians uttering the T-word (tax) has driven up the political stock of cap and trade over carbon tax, though it has been well argued that a tax is easier to administer and less likely to be vulnerable to financial gimmicks analogous to subprime loans as part of the deal. All of these important fine points will be debated over the next few years, both in the United States and the world at large.

Targets (caps) and timetables to achieve them were emphasized at Kyoto in 1997, in California since 2005, in other parts of the United States such as the Northeast, at the EU, and also in Australia in the state of South Australia and then nationally since their change in government in November 2007. What Australia will do remains uncertain, especially given the financial crisis.

Creating a shadow price—or a direct price via a carbon tax—is appropriate because without a tailpipe and smokestack dumping charge, people will (a) not work as hard to invent alternatives, and (b) emit with impunity. This is called the classic "externality" or "market failure." Since there's no cost to those companies for emitting, they therefore have no price incentives to cut emissions. To restore the market to its "optimal" state, the price of energy should reflect all the costs, including damages to nature and society from unpriced emissions.

Thus, governments must intervene to restore a true market by implementing rules that require that internalization of external costs like sending kids with asthma from air pollutants to the hospital or causing rising sea levels and coastal damages. But the issue is politically divisive. Wherever shadow prices have been proposed, even in California, which has a law on the books to require a stringent carbon emissions cap, the government is having a devil of a time working out the rules on how to do it in practice. Without doubt, a similar frenzy will follow the U.S. attempts in the Congress in 2009-2010 as a buildup toward the all-important Copenhagen UN Conference of the Parties meeting in December 2009. This meeting has been designated as the long-term follow up to the Kyoto Protocol of 1997, and it will likely be very contentious.

Europe has been calling for major caps. The United States under Obama has been agreeing in principle, but suggesting much smaller caps at first and then a ramp-up to major cuts a few decades ahead. That approach is what was needed to get the U.S. House of Representatives to pass the first climate control legislation in its history in June

2009, when the Waxman-Markey bill passed by a slim margin after months of rancorous debate and many last minute political trade-offs. The developing countries, including China and India, are saying they won't take cuts in emissions until they catch up per capita with the rich countries. Rich countries note that developing countries do indeed have lower per capita emissions levels, but with many times the populations of the rich countries, they made a choice for more people and lower per capita emissions. Their response, other than reminding the wealthy about their colonial occupations and market hegemony, accuses the rich of obscene overconsumption. The rich then note the corruption in governance in many developing countries followed by a typical response from the poorer nations that the rich have hoarded their consumption and wealth and have not helped with needed development strategies that lead to lower population growth rates. And it goes on and on, with point and counterpoint, recriminations and finger-pointing.

The only way out of this endless set of harangues is to fashion cooperative solutions, whereby rich countries first "walk the walk" by significantly cutting their own emissions. After all, the 20 percent or so of the world's population in the Organisation for Economic Co-operation and Development countries are responsible for about 75 percent of the accumulated historical emissions. Yet the developing world, especially India and China, will replace the rich as the big emitters of the future unless they avoid our model of wealth generation from the industrial revolution: sweatshops, dirty coal burning, and internal-combustion automobiles. We need to help the developing countries develop on a clean and green pathway by literally leapfrogging over the industrial revolution to high tech. This process has already happened in communications. The rich countries once strung copper wires all over their continents for phone lines, using lots of materials and energy. Now in the developing world they use cell phones and Skype.

We need to fashion cooperative solutions that can lead to leapfrogging in power production and transport systems—like renewable energy sources, smart grids to more efficiently distribute that energy, and much more energy efficiency in buildings and machines so less energy is required to keep the economy running. All of this will happen on its own, but ever so slowly relative to the critical need to curb emissions drastically. We need electric cars and properly produced biofuels that don't compete with food, as corn-based ethanol does in the United States.

All too often the C-word taught in business classes is "competition," but to solve the climate crisis the C-word we need more of is "cooperation." After World War I, the League of Nations was a miserable failure. But we did learn the lesson of ignoring cooperation, with the establishment of the United Nations in 1945 and the Marshall Plan in 1947. Thus, cooperation has happened before in response to wars and after emergencies like earthquakes and flooding catastrophes. This time we need cooperation in advance to prevent the new type of looming catastrophe. What kinds of steps on the right road? A price on carbon emissions is essential, but it is a time-consuming process, if European and Californian experiences are any guides. What are the earlier steps that are needed urgently now? Why don't we start with adaptation funds to help people through the unavoidable climate change in the pipeline?

Adapt to the Unavoidable. We must figure out what levels of climate change in each sector, region, or group that may be impacted is "dangerous"—a value judgment about what is unacceptable—for that system. Next, that assessment needs to be used to help define an urgent mitigation strategy to stay below those levels of likely nonadaptability. For some systems, like the Inuit hunting culture, Kiribati permanence, or the polar bear ecosystem, it is probably already too late to adapt successfully, so we will need to find ways to reduce impacts via alternative activities for the Inuit, new homes for some small islanders, or to fashion viable land-based habitats for polar bears. At the same time, these unavoidable impacts should be used to help motivate strong mitigation actions, so we don't end up having dozens of systems comparably damaged. But before mitigation is achieved via a price on carbon, there are other mitigation steps to implement even earlier—like now.

Performance Standards. One concrete step to take is creating performance standards for buildings and machines. Energy efficiency is the cheapest, best way to achieve this goal. California uses only 50 percent of the emissions and energy per capita of the U.S. average and is 300 percent better than Texas. The prime reason: California has a 35-year history of legally mandated performance standards like building codes, refrigerator and air conditioner standards, and so on. This policy, while initially contentious, is now popular with both Democrats and Republicans, because it saves the state 15 percent of its electricity bill annually, which for California is about $7 billion a year. That gets bipartisan attention in a positive way.

One reason California is so energy efficient relative to most of the rest of the United States is that the state uses a part of a gasoline tax to fund the California Energy Commission (CEC). The commission in turn examines the cost effectiveness of every appliance and building code. The CEC then tries to recommend mandatory standards that consider how long it takes to repay the investor or homeowner by saved energy bills. I like to use criteria I developed in 2006 in South Australia at the prompting of its premier, Michael Rann—what I call the "7/11 standard." If the extra cost of a more efficient refrigerator, for example, would be paid back in less than 11 years—equivalent to more than a 7 percent return on investment (about the average mortgage interest rate)—then the total monthly expenses for a typical family with a 7 percent or higher mortgage is actually lowered. In that case, such performance standards should be mandatory. I recall a minister in the Department of Treasury and Finance in South Australia chiding me, saying his state was "not a culture of mandatory."

I quipped, "But, Minister, when I drove over here I noticed three red-light cameras and one speed camera—those sound pretty mandatory to me!" He grimaced. "And imagine the carnage on the streets if traffic laws were guidelines and there were no cops and judges?" He got the point.

If an improvement has a reasonable payback, the performance standard should be mandatory. If it isn't, it's probably not going to happen. Texas doesn't have nearly as many mandatory performance standards as California. Therefore, they are very energy inefficient by comparison.

It can't simply be left to the states alone to deal with performance standards, since free riders can hurt national and global environmental amenities when their pollution blows past state borders. Broad-scale coverage of efficiency regulations are effective strategies both to reduce environmental stresses and to reduce the energy needed to produce our economic services, thus reducing our dependence on dirty or foreign sources of energy. Performance standards can offer the most immediate emissions cuts, and if reasonable payback criteria are followed, such "no regrets" policies can give environmental and security benefits at below zero cost (relative to the payback criteria). That's why they're popular in Europe, Japan, and other places that have strict codes for efficiency, with fair return criteria as their basis. We need national mandatory performance standards to be implemented as an urgent priority and thus help to reduce our emissions relative to business as usual *now*, not decades down the road when new technologies come on line to replace old polluting ones. The Waxman-Markey bill did recognize this need, as will most subsequent climate policy legislation, I am quite confident.

Incentives to Invent Our Way out of the Problem. Another step is to fashion incentives to invent our way out of the carbon emissions game—what I like to call a "learning-by-doing feeding frenzy." We have to get funding to the creative ideas in those hundreds of promising energy system start-ups out there. One day somebody's going to make billions of dollars when they invent a really efficient solar thermal system capable of energy storage, or produce biofuels with a negative CO_2 balance that isn't competing with food production. I can cite any number of worthy competitors—for instance, biochar or some other very promising ideas such as algae for biodiesel. But can we scale these up to the massive level that we need to replace the energy that produces a trillion tons of CO_2 over this century? That requires experimentation that leads to learning by doing. We have always done that. We did not start coal and nuclear power and the electronics industry by free market capitalism alone. Government subsidies and government-direct funding and government-determined friendly operating rules spurred fledgling industries. The Japanese still do that, and we will have to do the same for green technology developers.

To have those new clean technologies, we'll need investments. Does that mean loan guarantees? It could. At a recent conference in which I was advocating technology pump-priming via loan guarantees to venture capitalists and grants for promising ideas, a congressman who happened to be in the audience asked me, "How much do you think we'll need in loan guarantees to venture capitalists and the like to be able to have a significant impact on the rate at which we learn to get these cleaner technologies to scale?"

"Maybe $30 billion annually for a decade or so?" I answered.

"Thirty billion!" he roared. "Don't you read the newspapers? That's completely impossible. That's outrageous. We're having a major problem with the economy."

"But, Congressman, we just spent three-quarters of a trillion dollars in one year to bail out a bunch of underregulated, greedy bankers. Why can't we spend 4 percent of that every year for the next ten years or so to try to get planetary sustainability and long-term energy systems that will sustain the economy with growing numbers of jobs?" Our new President agrees and is working very hard to try to achieve that goal, despite partisan bickering in the Congress.

Put a Price on Carbon. The fourth step in the climate policy sequence, already described earlier in detail, is what many governments and the UN conferences are working on: cap and trade or carbon taxes. It is a critical component of effective climate policy, and it must happen if dangerous climatic impacts are to be reasonably reduced. But, as earlier said, given how long it will likely take to get it fully implemented, I am very leery of putting all our mitigation eggs in the cap and trade or carbon tax basket without also implementing immediately stronger performance standards and investment incentives. Each of these four steps mentioned so far is necessary, but none by itself is sufficient to effectively solve the problem.

Geoengineering. Many scenarios produce a rapidly multiplying set of interconnected risks and would undoubtedly spur calls for geoengineering schemes to try to offset the large impacts. This is made all the more acute by the threat of crossing many tipping points, such as a Greenland ice sheet's irreversible melting and many meters of eventual sea level rise. Geoengineering is not one-stop shopping. For instance, a cement process that actually takes CO_2 out of the air rather than giving it off, as in conventional cement manufacture, is being developed, and ideas are now advancing to invent efficient fuels manufactured from biomass wastes that both put some carbon back into nature and don't compete with food production. This concept is actually a CO_2 negative geoengineering technique called biochar. If biomass is cooked in a dry container at high temperature in a process known as pyrolosis, about half the carbon can be removed as an energy-rich gas, and much of the remaining solids becomes char—long-lived black elemental carbon. It improves the soil, so increased crop yields or plantation biomass could be grown in this cycle—and sequesters carbon for a millennium in soils as char. Another idea is to take CO_2 out of coalburning smoke stacks and put it underground or, alternatively, feed it to algae, which later can be used to make biodiesel fuel. These carbon capture and sequestration (CCS) processes are now at very early stages of development, but with a learning-by-doing feeding frenzy, as I like to say, and public investment to create incentives to venture capitalists to invest billions of private dollars, some of these now isolated start-ups could become major new megabusinesses in the business of doing well by doing good. In that sense this carbon removal type of geoengineering is a very positive development.

But the other kind of geoengineering—radiative forcing offsets, such as injecting dust into the stratosphere to offset the heat-trapping infrared radiation that results from increased levels of CO_2—has problems of miscalculation and planetary management. Of course, schemes such as injecting dust into the stratosphere to reflect light and offset radiative forcing from greenhouse gas buildup would not prevent potentially severe damage to the oceanic food webs associated with increasing oceanic acidification from ever increasing atmospheric concentrations of CO_2. And, since the increases in anthropogenic CO_2 concentration and associated warming are believed to last for a millennium or so,[5] the need for geoengineering management based on global international cooperation would have to be sustained without failure or interruption from wars (hot or cold) or political stresses when negative weather events occur. Moreover, the issue of liability to the climate controllers would inevitably be raised.[6]

So some forms of geoengineering might save the climate world, whereas others are doing the unthinkable. Geoengineering to cool the planet is a very last resort that we will be headed toward if we don't halt very dangerous climate change conventionally with performance standards, incentives for clean tech, and a price on carbon.

A WORD ABOUT TIPPING POINTS

People like to talk about tipping points. I, too, have used abrupt climatic impacts as a major reason for concern over higher warming levels. What's an example? Melting Greenland—once you have induced accelerating major deglaciation, a big chunk of the ice mass goes, at least for 10,000 years or so. I don't believe any single estimate is a threshold for dangerous climate change—we already have experienced some dangerous impacts in which climate change is partially implicated (for instance, Hurricane Katrina in 2005 and the 2003 killer heat wave in Europe were both likely intensified to some extent by warming).

What level of warming pushes us past some tipping point is not precisely known, given the many uncertainties that remain, but it's best described by a probability function—a bell curve. All you can say is the higher you go with warming above the present, the larger the number of systems that will be harmed and the more intensely they will be damaged. You would have to be crazy to take the risk that we will luck out on the bell curve and the threshold for Greenland deglaciation will be higher than anyone predicts—but that's risk management, not scientific "truth."[7]

Discussion of tipping points has become more prevalent in the media of late. What is new is the assertion of some folks that we know the level of warming where the tipping points are. For ice melt, it is freezing point, right? But not really, for radiative inputs that affect energy balance can melt ice when the temperature is below freezing or above—snow in sun melting on a below-freezing day or frost occurring on an evening warmer than freezing.

What is the tipping point for irreversible melting of Greenland where it won't stop once started, even if it takes centuries or more to play out? Well, the strict answer is—we don't know. But we have some pretty good ways to estimate with medium confidence (one-third to two-thirds chance) by looking at paleoclimates and recent ice sheet behavior. For Greenland to irreversibly melt, my own Bayesian priors would be roughly a 2 to 5 percent chance that it is already too late and it will happen over the long run. At 1 degree Celsius (1.8 degrees Fahrenheit) more warming, I'd raise the odds to 25 percent, and at 2 degrees Celsius (3.6 degrees Fahrenheit) to 60 percent, at 3 degrees Celsius (5.4 degrees Fahrenheit) to 90 percent, and so on. But while we wait for more consensus on these odds, we could be committing to dangerous tipping points, so precaution suggests, "Hedge early and often."

The EU's identification of 2 degrees Celsius (3.6 degrees Fahrenheit) warming above preindustrial temperatures as a "dangerous" threshold is as much a political judgment (we couldn't likely do much better than to keep warming to 2 degrees Celsius) as it is a scientific point—we start to accumulate many impacts at that level of warming. From the point of view of Inuit culture, polar bear ecosystem, and residents of Hurricane Alley, global warming is already getting dangerous.

I like to compare tipping points to a kids' skateboard park, where the ramp starts up slowly and gets nonlinearly steeper until it is vertical at the top, and the kid jumps and the parents hide their eyes! In other words, there are many such threshold tipping points in the bio-geophysical-social system, but the problem is we don't know precisely where they are. My skateboard ramp is an analogy to the steepening threats as we add warming. The warmer we get, the more systems are at risk and the deeper the impacts—the steeper the ramp in my analogy.

Tipping points matter, but too much specificity on levels for individual systems is a quick trip to loss of credibility. My usual framing is risk management. Why take major risks with the planetary life-support system when mitigating the risks can be done for a small fraction of the growth rate of the GDP and only a few years' delay in becoming 500 percent richer per capita over a century?[8] All that assumes we have a century-long average GDP growth rate of about 2 percent, which is less than in the first eight years of this century. Yet I have failed to popularize this point for over 15 years now, since I first debated it with Bill Nordhaus in 1992 and in a letter and response to *Science* in 1993. When I asked my friends in the media why they don't report this issue, they simply reply: "too many numbers."

TARGETING FOR THE FUTURE

The U.S. government stimulates science by deciding to whom to give research grants or other incentives. It also stimulates science because if the government says we really do want to have lower emissions, and we know that we have to have nanotube solar collectors or direct-current transmission lines—or whatever new, brilliant breakthrough that I'm hoping will happen—somebody has to fund it. After all, the object of government is to improve the quality of life of its citizens. Sometimes that involves targeting investments. We certainly target investments for the study of cancer and investments in national security. Why not target investments in sustainable development and a safer environment? A promising start by the Obama Administration, announced on June 23, 2009, by Secretary of Energy Stephen Chu, includes three loans to automakers Tesla Motors, Nissan, and Ford, the first awarded under the $25 billion Advanced Technology Vehicle Manufacturing Program to help automakers offset the cost of retooling to build eco-friendlier cars that are at least 25 percent more fuel efficient than 2005 models.

My old friend John Holdren, now serving as the director of the White House's Office of Science and Technology Policy (S&T), is in the right place and at the right time to guide the setting of targets for the United States. He wrote an editorial in May 2009, published in *Science,* in which he lays out a solid plan. Even just a few paragraphs demonstrate the Obama Administration's commitment:

> I see the top S&T priorities for the Obama administration in terms of four practical challenges and four cross-cutting foundations of

success in addressing all of them. The practical challenges are: bringing S&T more fully to bear on driving economic recovery, job creation, and growth; driving the energy-technology innovation needed to reduce energy imports and climate-change risks while creating green jobs and competitive new businesses; applying advances in biomedical science and information technology together to help Americans live longer, healthier lives with reduced health care costs; and ensuring that we have the defense, homeland security, and national intelligence technologies needed to protect our troops, citizens, and national interests, and to verify the old and new arms control and non-proliferation agreements that are likewise essential to our security.

The cross-cutting foundations of success are: increasing the capacities and output of our country's fundamental research institutions, including our great research universities and major public and private laboratories and research centers; strengthening science, technology, engineering, and mathematics education at every level, from precollege to postgraduate to lifelong learning; improving and protecting the information, communication, and transportation infrastructures that are essential to our commerce, science, and security alike; and maintaining and vigorously exploiting a cutting-edge set of capabilities in space, which must be understood not just as grand adventure and focus for expanding our knowledge of how the universe works, but also as a driver of innovation and a linchpin of communications, geopositioning technology, intelligence gathering, and Earth observation.

It is a lot to get done. But led by a President who deeply grasps the importance of S&T to our national goals and who is putting scientists, engineers, and innovators back into the center of what the executive branch does, "Yes, we can." [9]

I was in Stockholm, Sweden, in late May 2009 and met for 90 minutes with those responsible for climate policy in the prime minister's office. Sweden will hold the presidency of the EU when the Copenhagen meeting takes place in mid-December 2009. They want to broker a strong deal on their watch. They told me that they are very impressed with the turnaround in the United States toward responsible climate policy and trust the Obama Administration to keep its word on big cuts later in the century. "But what about your Congress?" I was asked. "Will the Senate get the 60 votes to pass the climate legislation?"

"Probably," I said, "since Congress would rather have a bill that pleases nobody than to have the EPA regulators do the job without congressional input."

"Fine," they said, "but if we get a treaty done, and Obama signs it like Clinton did on Kyoto, does that mean the Senate can muster 67 votes to pass the Copenhagen treaty?"

"Good question," I replied, complimenting them on their knowledge of the American political machinery. "It will be a close call. But remember, when the U.S. President signed the Strategic Arms Limitation Treaty but Congress refused to ratify, each subsequent administration has followed it anyway."

"What about aid to developing countries for adaptation or to help them leapfrog to the clean technology that you are always advocating? Will the Congress support that?"

"That's a tough call too. It would be a hard sell on Capitol Hill to just give away billions in some donor-receiver mode. Much better would be to set up joint ventures where American or Swedish or Japanese companies could work with Indian or Chinese counterparts to build green tech over there. Profits, patents, et cetera, should be shared, so what is learned by this investment could be brought back to the developed country partners. That way it would be a win-win framing, not donor-receiver, which is politically hard to sell to donor nations' citizens and is often demeaning to recipient nations being told what to do with their donations. Partnerships and win-win situations are the key for such intellectual and financial transfers to developing countries. We need to think outside of the box on this to have a political home in developed or developing countries. But of course this is just my reading of the political tea leaves now."

"What we really are trying to learn from you," the Swedes admitted, "is whether the U.S. will be a reliable partner with us, or will they, like several times before, refuse to ratify treaties they signed because the Senate was more reluctant for international cooperation than the President?"

"All I can say is that Obama is very popular now, and this may be the best possible time for international cooperative agreements. Just understand that the less the U.S. has to do in a screaming hurry, the higher the chances for a good deal over the long haul. We have to have a generational perspective on this, although we can't wait a generation to get it started. That has to be in Copenhagen this year!" We all agreed on that and promised to work our side of the street as best we could.

Looking toward the future in the United States, I am optimistic and yet I want to remain realistic about the possibilities. Reducing the level of CO_2 emissions down to 1990 levels in the country as a whole by 2020 will be tough, and to go well below that is virtually impossible, given realpolitik in the United States and the generation it will take to replace major infrastructure.[10] The Waxman-Markey bill only argues a cut to 17 percent of 2005 emissions by 2020, not of 1990 emissions, which were some 15 percent lower than 2005—but this goal accelerates rapidly over time (83 percent cut in 2050). As I've said before, aspirational goals based on avoiding dangerous climate change are important to articulate, but we cannot expect that coal power plants will simply be shut down well before their economic lifetime expires or that we will build the new infrastructure—solar and wind and efficient "smart grid"—in a decade. Most important to me is that we don't waste the next decade with Band-Aids to get the numbers down initially and thus miss the opportunity to invest heavily in long-term solutions and invent our way out of this mess more permanently. I am leery of more "targets without teeth," as we have seen in the past.

It is very likely that emissions globally will continue to increase, with a peak in emissions and probably in concentrations of greenhouse gases in a few decades. This overshoot in concentration will, we hope, be followed by a steep decline in emissions and (with a decade or two lag) in concentrations, toward a much safer, low-stabilization level by century's end, if not a few decades sooner. So investment and deployment of low-emitting technology at a massive scale over the next two decades, along with not replacing bad emitters with the same or similar high-emitting technology, will be essential.

Other key elements include energy efficiency regulations, adaptation strategies for climate change unavoidably in the pipeline and a shadow price on carbon, and more smart growth planning and vehicles on a diet. And despite all those promising prospects for deep cuts later in the century, we are still likely to have several very risky decades in the next 40 years—we might call it the "overshoot hazard period"—where we temporarily exceed thresholds that most climate impact specialists would likely label as "very risky" in their risk-management judgment. We must study carefully what such an overshoot period would imply for planetary risks, but even while that research is going on, we still have to work very hard to keep the risky peak as low as possible, have it occur as soon as possible, and have the peak drop off toward safer stabilization levels as fast as possible. To achieve that will take research, development, *and* deployment of many clean technologies and sensible rules for efficient use of energy production and end uses, and a "polluter pays" price on carbon.

So that is my view. Some groups are calling for emissions cuts to 25 percent below 1990 levels by a decade from now. Would that it were possible. Barring a dramatic breakthrough in both costs of alternative technologies and political will, big cuts below 1990 levels of emissions on a global scale by 2020 are not even in the ballpark of feasibility, although I'll fight and hope we do better than a continuing increase over the next ten years. I do genuinely believe we will slow down the rate of emissions growth immediately, reach the peak in emissions in the United States ideally before 2020, and start to slow concentration buildups and eventually have that peak followed by steep declines by 2030. But that prediction is just for the rich countries.

Developing countries—unless we set up strong cooperative ventures with them to leapfrog over the old technologies—will elevate the planet's concentration levels to much more dangerous heights. But with cooperation we can bring that down substantially too—given political will and joint solutions. So there will be plenty more climate change built into the pipeline, I fear, despite the promising turnaround in the growth in emissions in the rich countries. The amount and rate of globally abated greenhouse gas emissions is still likely to be much less than what we need to stay below the 2-degree-Celsius (3.6-degree-Fahrenheit) warming target. "Dangerous anthropogenic interference with the climate system" on a large scale is simply getting uncomfortably close.

Meanwhile, as the business leviathans, environmental heavies, and governmental moguls wrestle at Copenhagen and other capitals, each of us can do our part. Planning trips to minimize automobile use to what is necessary; conserving energy at home; buying energy efficient appliances when we shop; using less imported foodstuffs, and eating more local foods, especially fruits and vegetables are all under our control. Buying efficient cars—I am waiting a few years to replace our standard hybrid with a plug-in variety—is something we can consciously choose to do.

Form a reading group with friends and neighbors to be better informed and, as I've shown over and over again in these pages, less susceptible to the rampant propaganda that is so prevalent on climate issues. Link with those groups that seem credible to you, and show up at city council meetings to insist your town goes greener. Be sure to support local politicians who stand up for sustainability, because they need your support. Get green towns to cooperate and pressure states to green up. It all can be done without your having the authority to negotiate with the Chinese.

But most important, for me, as grandparent, parent, and teacher, is to hum in your head often the lines of the Crosby, Stills, and Nash song from decades ago. The advice is still the most important thing any of us can do as individuals: "Teach your children well."

ACKNOWLEDGMENTS

NO ACCOUNTING OF A FOUR-DECADE STORY is the sole work of one witness. My journey was only possible because of mentors and colleagues, many of whom have already been featured in these pages. Prominent among the very many for their positive influences on me are Ken Arrow, Jesse Ausubel, Christian Azar, Wilfred Bach, Eric Baron, Andy Beattie, Andre Berger, Wolf Berger, Bert Bolin, Penny Boston, Elise Boulding, Wally Broecker, John Brockman, Mikhail Budyko, Phil Campbell, Bob Chen, Bob Chervin, C. K. "John" Chu, Ralph Cicerone, Bill Clark, Jim Coakley, Curt Covey, Paul Crutzen, Gretchen Daily, Jared Diamond, Bob Dickinson, Kris Ebi, Jack Eddy, Paul Edwards, Paul and Anne Ehrlich, Chris Field, John Firor, Inez Fung, Tzvi Gal-Chen, Larry Gates, Mickey Glantz, Al Gore, Randi Londer Gould, Larry Goulder, Jim Hansen, John Harte, Danny Harvey, Klaus Hasslemann, David Hawkins, Teresa Heinz, Ann Henderson-Sellers, John and Cheri Holdren, Chris Hotham, Malcolm Hughes, Jill Jaeger, Phil Jones, Tom Karl, John Katzenberger, William Kellogg, Don Kennedy, Don Kinney, Jeff Koseff, John Kutzbach, Hubert Lamb, Simon Levin, Diana Liverman, Ed Lorenz, Tom Lovejoy, Jim Lovelock, Amory Lovins, Hunter Lovins, Jane Lubchenco, Linda Mearns, Jerry Mahlman, Suki Manabe, Mike MacCracken, Tom Malone, Mike Mann, Lynn Margulis, Cliff Mass, Mike Mastrandrea, Pam Matson, Margaret Mead, Jim Miller, Hal Mooney, Granger Morgan, Richard Moss, Naki Nakicienovic,

Jerry Namias, Bill Nordhaus, Jerry North, Mike Oppenheimer, Naomi Oreskes, Rajenda Pachauri, Graeme Pearman, John Perry, Rick Piltz, Stuart Pimm, Barrie Pittock, Jeff Price, Ron Prinn, Stefan Rahmstorf, V. Ramanathan, S. Ichtiaque Rasool, Roger Revelle, Andy Revkin, Mary Rickel, Walt Roberts, Alan Robock, Norm Rosenberg, Bill Ruddman, Carl Sagan, Jim Salinger, Ben Santer, John Schellnhuber, Mike Schlesinger, Joe Smagorinsky, Joel Smith, Kirk Smith, Susan Solomon, Richard Somerville, Annamaria Talas, Phil Thompson, Starley Thompson, Kevin Trenbeth, Rich Turco, Billie Turner, Peter Vitousek, Jean-Pascal van Ypersele, Bud Ward, Steve Warren, Warren Washington, Bob Watson, John Weyant, Bob White, Tom Wigley, Robyn Williams, Tim and Wren Wirth, George Woodwell, Gary Yohe, and many more.

For the production of this book, I gratefully acknowledge the professional skills for writing and editing help from Maura Shaw, my National Geographic Society editor Lisa Thomas, John Paine, Mary Beth Keegan, and Jane Sunderland, and the insights and efforts of literary agent Carol Susan Roth in helping me focus the concept and arrange for a great publisher for the manuscript. Sarah Jo Chadwick, my administrative assistant, provided immeasurable assistance as did Patricia Mastrandrea and Katarina Kivel. I also appreciate the comments on early draft manuscripts from many of those already listed, as well as Riley Dunlap, and Josh Howe. I thank Tim Flannery for kindly writing the Foreword and Erik Rasmussen for adopting the book for the Copenhagen Climate Council's work at the 2009 Copenhagen meetings.

Finally, no successful effort can be accomplished without the emotional support of my family, Becca Cherba and Adam Schneider, and my life partner, frequent co-author, fellow IPCC teammate, and most honest critic, Terry Root. Without her I probably wouldn't have survived my bout with cancer in 2002 or been nearly as effective in making our case for conserving the climate and the coupled human-natural systems that depend on it.

NOTES

INTRODUCTION

1. NASA/JPL, image PIA11194: "Global Carbon Dioxide Transport from AIRS Data, July 2008." Available online at http://photojournal.jpl.nasa.gov/catalog/PIA11194.

CHAPTER 1

1. Roger Revelle and Hans E. Suess, "Carbon Dioxide Exchange Between Atmosphere and Ocean and the Question of an Increase of Atmospheric CO_2 During the Past Decades," *Tellus* 9 (1957): 18-27.

2. George Will, "Al Gore's Green Guilt," *Washington Post*, September 3, 1996, a23, quoting Christopher M. Byron in an undated *New York* magazine article. Will's comment reads: "Gore is marching with many people who not long ago were marching in the opposite direction. New York magazine's Christopher Byron notes that Stephen Schneider of the National Center for Atmospheric Research in Colorado, is an 'environmentalist for all temperatures.' Today Schneider is hot about global warming; 16 years ago he was exercised about global cooling. There are a lot like him among today's panic-mongers." See my rebuttal; "Hot About Global Warming," *Washington Post*, September 26, 1992.

3. For an independent assessment, see Thomas C. Peterson, William M. Connolley, and John Fleck, "The Myth of the 1970s Global Cooling Scientific Consensus," *Bulletin of the American Meteorological Society* (September 2008): 1325-1337, especially Table 1, 1332.

4. S. Ichtiaque Rasool and Stephen H. Schneider, "Atmospheric Carbon Dioxide and Aerosols: Effects of Large Increases on Global Climate," *Science* 173 (1971): 138-141.

5. Prominent among these were my soon-to-be colleagues Warren Washington and Robert Dickinson at the National Center for Atmospheric Research (NCAR) in Boulder, Colorado; Jim Hansen and Richard Somerville at GISS; Syukuro (Suki) Manabe at the Geophysical Fluid Dynamics Laboratory in Princeton; Tom Wigley at the University of East Anglia in the U.K.; and Larry Gates and Mike Schlesinger, then at Rand Corporation in Santa Monica, California.

6. G. Yamamoto and M. Tanaka, "Increase of Global Albedo Due to Air Pollution," *Journal of Atmospheric Sciences* 29 (1972): 1405-1412.
7. Carroll M. Wilson and William H. Matthews, eds., *Inadvertent Climate Modification: Report of the Study of Man's Impact on Climate (SMIC)* (Cambridge: MIT Press, 1971).

CHAPTER 2
1. When I introduced Jim Hansen at a lecture recently at Stanford and recounted this little tidbit, he claimed he couldn't remember ever saying that. So it's my memory against his—probably we're both wrong after 40 years!
2. Among them were Bob Dickinson, John Mitchell from the United Kingdom, Mike Schlesinger (then at Rand), Mike MacCracken (then at Lawrence Livermore National Laboratory), Jerry North from St. Louis, and others too numerous to include.
3. Stephen H. Schneider, "Hot About Global Warming," *Washington Post*, September 26, 1992.
4. Stephen H. Schneider, with Janica Lane, *The Patient From Hell: How I Worked With My Doctors to Get the Best of Modern Medicine and How You Can Too* (New York: Da Capo Press, 2005).

CHAPTER 3
1. World Meteorological Organization, *Proceedings of the World Climate Conference: A Conference of Experts on Climate and Mankind*, Geneva, February 12-22, 1979. Publication No. 537.
2. Ibid.
3. July 30, 1979, testimony before the U.S. Senate, Committee on Governmental Affairs (Senator Abraham Ribicoff, Chairman), Carbon Dioxide Accumulation in the Atmosphere, Synthetic Fuels and Energy Policy, a Symposium (Washington, D.C.: U.S. Government Printing Office).
4. The following material is quoted from July 31, 1981, testimony in "Carbon Dioxide and Climate: The Greenhouse Effect," Hearing Before the Subcommittee on Natural Resources, Agriculture Research and Environment and the Subcommittee on Investigations and Oversight of the Committee on Science and Technology, U.S. House of Representatives, 97th Congress (Washington, D.C.: U.S. Government Printing Office).
5. Robert Reinhold, "Reagan Aides Fail to Mollify Worried Scientists," *New York Times*, October 27, 1981.
6. John F. Kennedy, speech given in Tulsa, Oklahoma, September 16, 1959, printed in John F. Kennedy, *The Strategy of Peace*, edited by Allan Nevins (New York: HarperCollins, 1960).

CHAPTER 4
1. *Chemical Weekly*, July 16, 1975.
2. Several others were also contemporary pioneers, including Ralph Cicerone, Harold Johnston at University of California–Berkeley, and Mike McElroy at Harvard.

3. James Hansen, quoted in Kate Sheppard, "A Voice in the Wilderness," *The Guardian*, June 23, 2008. See also the discussion in my book *Global Warming: Are We Entering the Greenhouse Century?* (San Francisco: Sierra Club Books, 1989).

4. "National Energy Policy Act of 1988 and Global Warming: Hearings Before the Committee on Energy and Natural Resources, United States Senate, One Hundredth Congress, second session, on S. 2667... August 11, September 19 and 20, 1988."

5. James Hansen speaking at the National Press Club, Washington, D.C., on June 23, 2008.

6. Ross Gelbspan, *The Heat Is On: The Climate Crisis, the Cover-up, the Prescription* (New York: Basic Books, 1998); Dave Gilsen, "Hot and Bothered: An Interview with Ross Gelbspan," *The Paragraph*, April 18, 2005. Available online at http://theparagraph.com/hot-and-bothered-an-interview-with-ross-gelbspan/.

7. Aaron Swartz, "Raw Thought: Shifting the Terms of Debate: How Big Business Covered Up Global Warming." Available online at http://www.aaronsw.com/weblog/shifting1.

8. Aaron M. McCright and Riley E. Dunlap, "Defeating Kyoto: The Conservative Movement's Impact on U.S. Climate Change Policy," *Social Problems* 50, No. 3 (2003): 348-373.

9. "Editors' Note," *New York Times*, May 2, 2009, appended to Andrew C. Revkin, "A Climate of Doubt," *New York Times*, Dot Earth blog, April 23, 2009; Andrew C. Revkin, "Industry Ignored Its Scientists on Climate," *New York Times*, Dot Earth blog, April 24, 2009.

10. Shardul Agrawala, "Context and Early Origins of the Intergovernmental Panel on Climate Change," *Climatic Change* 39 (1998): 605-620.

CHAPTER 5

1. Severn Cullis-Suzuki's speech at the 1992 Rio Earth Summit can be viewed on YouTube.com: http://www.youtube.com/watch?v=TQmz6Rbpnu0.

2. *IPCC Second Assessment: Climate Change 1995: A Report of the Intergovernmental Panel on Climate Change* (Geneva: IPCC, 1996), Chapter 8, Section 4, 22.

3. Frederick Seitz, "A Major Deception on Global Warming," op-ed in *Wall Street Journal*, June 12, 1996.

4. Ibid.

5. Ben Santer et al., letter to the *Wall Street Journal*, June 25, 1996.

6. Bert Bolin, John Houghton, and Luiz Gylvan Meira Filho, letter to the *Wall Street Journal*, June 25, 1996.

7. S. Fred Singer, "Coverup in the Greenhouse?" *Wall Street Journal*, July 11, 1996.

8. P. N. Edwards and S. H. Schneider, "Self-Governance and Peer Review in Science-for-Policy: The Case of the IPCC Second Assessment Report," in *Changing the Atmosphere: Expert Knowledge and Environmental Governance*, edited by Clark Miller and Paul Edwards (Cambridge: MIT Press, 2001), 219-246.

9. Letter from the Executive Committee of the American Meteorological Society and the Trustees of the University Corporation for Atmospheric Research, July 25, 1996, UCAR quarterly newsletter (Summer 1996).

10. For an excellent account of this story and the petition's revival in 2007, see the Oregon Petition in the Scientific Consensus section on the website "The Global Warming Debate: A Layman's Guide to the Science and Controversy." Available online at cce.890m.com/scientific-consensus/.

11. Adapted from Stephen H. Schneider, "Kyoto Protocol: The Unfinished Agenda," editorial essay, *Climatic Change* 39 (1998): 1-21.

12. A. Graumann, N. Lott, S. McCown, and T. Ross, "Climatic Extremes of the Summer of 1998," National Climatic Data Center Technical Report No. 98-03, November 1998 (Asheville, NC: National Climatic Data Center, 1998).

13. Some portions of this chapter are adapted from Stephen H. Schneider, with Janica Lane, *The Patient From Hell: How I Worked With My Doctors to Get the Best of Modern Medicine and How You Can Too* (New York: DaCapo Press, 2005).

14. UN Environment Programme and the Center for Clouds, Chemistry and Climate, Impact Study, "The Asian Brown Cloud: Climate and Other Environmental Impacts," August 2002.

CHAPTER 6

1. Bill Maher, HBO television program *Real Time*. Available online at http://www.youtube.com/watch?v=H9mWZZ2U6EQ.

2. Eric Berger, "Hurricane Experts Reconsider Global Warming's Impact," *Houston Chronicle*, April 12, 2008.

3. Passages from the Fourth Assessment Report Chapter 19 are taken from S. H. Schneider, et al., "Assessing Key Vulnerabilities and the Risk from Climate Change," in *Climate Change 2007: Impacts, Adaptation and Vulnerability. Contribution of Working Group II to the Fourth Assessment Report of the Intergovernmental Panel on Climate Change*, edited by M.M. L. Parry, O. F. Canziani, J. P. Palutikof, P. J. van der Linden, and C. E. Hanson, eds. (Cambridge: Cambridge University Press, 2008), 779-810.

4. "Burning Embers" chart, published in Joel B. Smith et al., "Assessing Dangerous Climate Change Through an Update of the Intergovernmental Panel on Climate Change (IPCC) 'Reasons for Concern,'" *Proceedings of the National Academy of Sciences* 106 (2009): 4133-4137.

CHAPTER 7

1. Eric Boehlert and Jamison Foser, "Flouting Scientific Opinion, Stossel Promoted Michael Crichton's Global Warming Skepticism," December 16, 2004 blog entry, Media Matters for America website. Available online at http://mediamatters.org/items/200412170002.

2. Andrew C. Revkin, "CO_2 = Pollution. Now What?" *New York Times*, April 17, 2009, published on the website.

3. Andrew C. Revkin, "Climate Experts Tussle over Details. Public Gets Whiplash." From the *New York Times*, July 29, 2008 © 2008 The New York Times. All rights reserved. Used by permission and protected by the Copyright Laws of the United

States. The printing, copying, redistribution, or retransmission of the Material without express written permission is prohibited.

See also Revkin's July 29, 2008, *New York Times* Dot Earth blog, "Whiplash Effect and Greenhouse Effect."

4. Revkin, "Climate Experts Tussle."

5. Jonathan Schell, "Our Fragile Earth," *Discover* (October 1989), 47.

6. "Loads of Media Coverage," *Detroit News* editorial, November 22, 1989.

7. "Defending Science," *The Economist*, January 31, 2002.

8. Charles Krauthammer, "Global Warming Fundamentalists: This Is Nuclear Winter Without the Nukes," *Washington Post*, December 5, 1997.

9. Ibid.

10. Stephen Schneider, "Twisted Revision," *Washington Post*, January 9, 1998.

11. Matthew Mosk and Juliet Eilperin, "Palin Not Convinced on Global Warming," *Washington Post*, August 29, 2008, published online.

12. See Lone Frank, "Scientific Conduct: Charges Don't Stick to The Skeptical Environmentalist," *Science*, January 2, 2004, 28b, and a Reuters article entitled "Panel: Danish Environmentalist Work 'Unscientific' " pertaining to a similar report on Lomborg performed by a panel of Scandinavian scientists (August 26, 2003).

13. Bjørn Lomborg, "Don't Waste Time Cutting Emissions," op-ed article, *New York Times*, April 25, 2009, published on the website. For an interesting Grist.com series on communicating with climate skeptics, see Coby Beck, "How to Talk to a Climate Skeptic: Responses to the Most Common Skeptical Arguments on Global Warming." Available at www.grist.org/article/series/skeptics/.

14. This ABC special program can be found in several segments on YouTube. Available at www.youtube.com.

15. Ibid.

16. George Monbiot, "We All Make Mistakes but George Will Just Won't Admit His," George Monbiot's blog in the *Guardian*, March 3, 2009, published on the *Guardian* website.

CHAPTER 8

1. S. H. Schneider, S. Semenov, A. Patwardhan, I. Burton, C. H. D. Magadza, M. Oppenheimer, A. B. Pittock, A. Rahman, J. B. Smith, A. Suarez, and F. Yamin, "Assessing Key Vulnerabilities and the Risk from Climate Change," in *Climate Change 2007: Impacts, Adaptation and Vulnerability. Contribution of Working Group II to the Fourth Assessment Report of the Intergovernmental Panel on Climate Change*, edited by M. M. L. Parry, O. F. Canziani, J. P. Palutikof, P. J. van der Linden, and C. E. Hanson (Cambridge: Cambridge University Press, 2008), 779-810.

2. Adapted from *Nature* 458, (April 29, 2009): 1104-1105.

3. Nicholas Schmidle, "Wanted: A New Home for My Country," *New York Times*, May 10, 2009.

4. Elisabeth Rosenthal, "Rich-Poor Divides Still Stall Climate Accord," *New York Times* Dot Earth blog, April 10, 2009.

5. Alaska Superstation: Anchorage, Fairbanks, and Juneau, news report, April 25, 2009,

published online at www.aksuperatation.com; Kimberley D. Mok, "Indigenous People's Climate Change Summit Giving 'Unified Voice,' " published online at www.treehugger.com.

6. Adapted from Carroll Harrington and Loma Fear, "Questions and Answers: Conversations with Steve Schneider," *Bay Area Green*, January, February, March 2009.

7. Stephen H. Schneider, Terry L. Root, and Patricia R. Mastrandrea, "Climate Change and Wild Species," in *Encyclopedia of Biodiversity* (Amsterdam and New York: Elsevier, 2007), 11-12.

8. Andrew C. Revkin, "Bush Aide Softened Greenhouse Gas Links to Global Warming," *New York Times*, June 8, 2005.

9. "Face of NIWA Sacked for Talking to Media," *ONE News*, April 24, 2009, published online at www.tvnz.co.nz.

CHAPTER 9

1. July 30, 1979, testimony by Stephen H. Schneider before the U.S. Senate, Committee on Governmental Affairs (Senator Abraham Ribicoff, Chairman), Carbon Dioxide Accumulation in the Atmosphere, Synthetic Fuels and Energy Policy, a Symposium (Washington, D.C.: U.S. Government Printing Office), 11.

2. February 28, 2007, testimony by Stephen H. Schneider to the Committee on Ways and Means, U.S. House of Representatives (The Honorable Charles B. Rangel, Chairman), Climate Change Risks and Control Strategies, 4.

3. Historian Naomi Oreskes of the University of California–San Diego Science Studies Program reported that several of the more aggressive climate contrarians such as Fred Singer and Fred Seitz had been on committees and boards of the Tobacco Institute or were consultants for tobacco companies. Naomi Oreskes, lecture, "The American Denial of Global Warming." Available online at http://www.climatesciencewatch.org/index.php/csw/details/oreskes_lecture/.

4. Gary Graff, "Dave Matthews Band Leads Epic Saturday at Rothbury," *Billboard*, July 6, 2008.

5. S. Solomon, G.-K. Plattner, R. Knutti, and P. Friedlingstein, "Irreversible Climate Change Due to Carbon Dioxide Emissions," *Proceedings of the National Academy of Science* 106 (2009): 1704-09.

6. Stephen H. Schneider, "Geoengineering: Could We or Should We Make It Work?" *Philosophical Transactions of the Royal Society A* 366 (Nov. 13, 2008): 3843-3862.

7. Adapted from Stephen Schneider, "Modeling the Future: A Talk with Stephen Schneider," interviewed by Russell Weinberger, *Edge* 241 (April 1, 2008), published online at www.edge.org.

8. C. Azar and S. H. Schneider, "Are the Economic Costs of Stabilizing the Atmosphere Prohibitive?" *Ecological Economics* 42 (2002): 73-80.

9. John P. Holdren, "Science in the White House," *Science* 324 (May 1, 2009): 567.

INDEX